VOL.1

Uncovering
Student Ideas
in **Primary Science**

25 New Formative Assessment Probes for Grades K–2

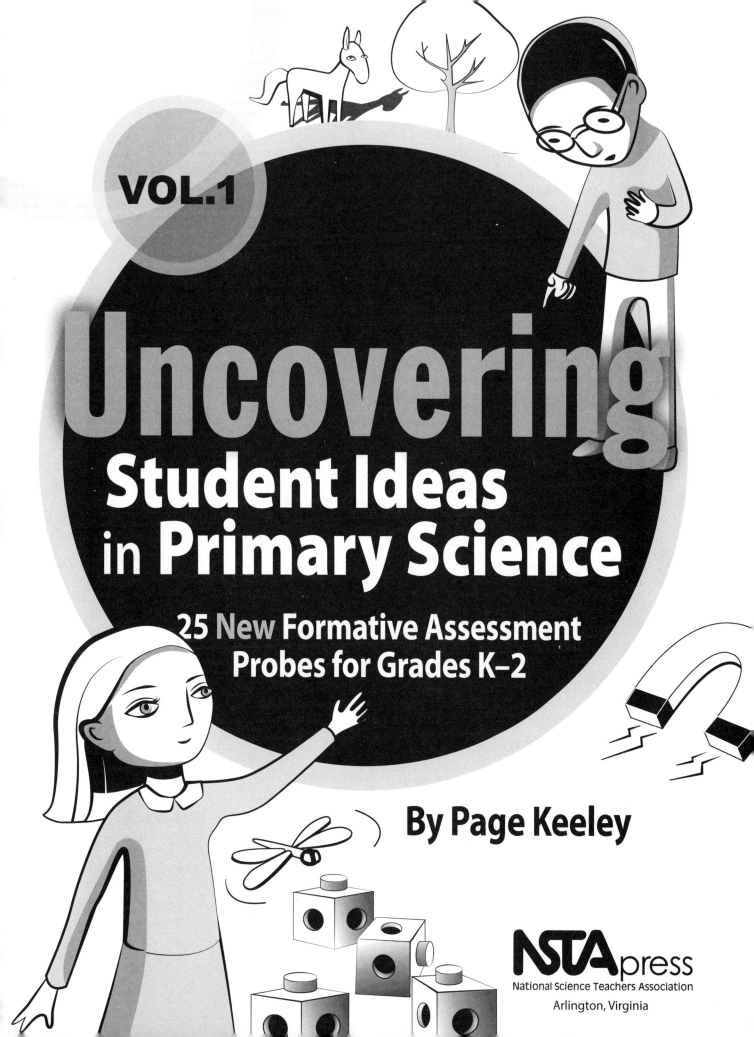

VOL.1

Uncovering
Student Ideas
in Primary Science

25 New Formative Assessment Probes for Grades K–2

By Page Keeley

NSTApress
National Science Teachers Association
Arlington, Virginia

National Science Teachers Association

Claire Reinburg, Director
Jennifer Horak, Managing Editor
Andrew Cooke, Senior Editor
Amanda O'Brien, Associate Editor
Wendy Rubin, Associate Editor
Amy America, Book Acquisitions Coordinator

ART AND DESIGN
Will Thomas Jr., Director
Cover, Inside Design, and Illustrations by Linda Olliver

PRINTING AND PRODUCTION
Catherine Lorrain, Director
Nguyet Tran, Assistant Production Manager

NATIONAL SCIENCE TEACHERS ASSOCIATION
David L. Evans, Executive Director
David Beacom, Publisher

1840 Wilson Blvd., Arlington, VA 22201
www.nsta.org/store
For customer service inquiries, please call 800-277-5300.

Copyright © 2013 by the National Science Teachers Association.
All rights reserved. Printed in the United States of America.
20 19 18 17   6 5 4 3

FSC
www.fsc.org
MIX
Paper from
responsible sources
FSC® C011935

PERMISSIONS
Book purchasers may photocopy, print, or e-mail up to five copies of an NSTA book chapter for personal use only; this does not include display or promotional use. Elementary, middle, and high school teachers may reproduce forms, sample documents, and single NSTA book chapters needed for classroom or noncommercial, professional-development use only. E-book buyers may download files to multiple personal devices but are prohibited from posting the files to third-party servers or websites, or from passing files to non-buyers. For additional permission to photocopy or use material electronically from this NSTA Press book, please contact the Copyright Clearance Center (CCC) (*www.copyright.com*; 978-750-8400). Please access *www.nsta.org/permissions* for further information about NSTA's rights and permissions policies.

Library of Congress Cataloging-in-Publication Data
Keeley, Page.
  Uncovering student ideas in primary science / by Page Keeley.
    volumes ; cm. — (Uncovering student ideas in science series ; 8-)
  Includes bibliographical references and index.
  ISBN 978-1-936959-51-8 -- ISBN 978-1-938946-87-5 (e-book) 1. Science--Study and teaching (Early childhood) 2. Educational evaluation. 3. Early childhood education—Aims and objectives. I. Title.
  LB1139.5.S35K44 2013
  372.35'044—dc23
                                    2013020706

Cataloging-in-Publication Data are also available from the Library of Congress for the e-book.
e-LCCN: 2013024236

Contents

Section 3. Earth and Space Science

Dedication

This book is dedicated to Emma Elizabeth Keeley, Lincoln Wright DeKoster, Court Wilson Brown, Jack Anthony Morgan, Cadence Jane Friend-Gray, and Emeline Leslie Friend-Gray. May you all grow up to have wonderful ideas!

Preface

This is the eighth book in the *Uncovering Student Ideas in Science* series, and the first one that exclusively targets young children's ideas. Like its predecessors, this book provides a collection of formative assessment probes designed to uncover the ideas students bring to their science learning. Each probe is carefully researched to elicit commonly held ideas young children have about phenomena or scientific concepts. A best answer is provided along with distractors designed to reveal research-identified misconceptions held by young children.

A major difference between this book and others in the *Uncovering Student Ideas in Science* series lies in the format of the student pages. The probes in this book use minimal text so that they can be used with children who are just developing their reading and writing skills. Each probe provides a visual representation of the elicited idea using familiar phenomena, objects, and organisms or set in situations that can be duplicated in the classroom. For example, "Is It Living?" elicits students' ideas about living and nonliving things using pictures of familiar objects and organisms. "Big and Small Magnets" uses a concept cartoon format to elicit children's ideas about magnetism, which can then be tested in the classroom using magnets of different sizes. The visuals are designed to capture children's interest and stimulate their thinking. Each probe ends by asking, "What are you thinking?" to draw out students' reasons for their answer choices and encourage "science talk."

Other *Uncovering Student Ideas* Books That Include K–2 Probes
While this book is specifically designed for K–2 students, other books in the series include K–12 probes that can be used or modified for the primary grades. The following is a description of each of the other books in the *Uncovering Student Ideas in Science* series and selected probes that can be used as is or modified for the primary grades:

Uncovering Student Ideas in Science, Volume 1 (Keeley, Eberle, and Farrin 2005)
The first book in the series contains 25 formative assessment probes in life, physical, and Earth and space science. The introductory chapter provides an overview of what formative assessment is and how it is used. Probes from this book that can be used in grades K–2 include:

- "Making Sound"
- "Cookie Crumbles"
- "Is It Matter?" (This probe has been modified for this book.)
- "Is It an Animal?" (This probe has been modified for this book.)
- "Is It Living?" (This probe has been modified for this book.)
- "Wet Jeans"

Uncovering Student Ideas in Science, Volume 2 (Keeley, Eberle, and Tugel 2007)
The second book in the series contains 25 more formative assessment probes in life, physical, and Earth and space science. The introductory chapter of this book describes the link between formative assessment and instruction. Probes from this book that can be used in grades K–2 include:

Preface

- "Is It a Plant?" (This probe has been modified for this book.)
- "Needs of Seeds"
- "Is It a Rock?" (version 1)
- "Is It a Rock?" (version 2)
- "Objects in the Sky"

Uncovering Student Ideas in Science, Volume 3 (Keeley, Eberle, and Dorsey 2008)

The third book in the series contains 22 formative assessment probes in life, physical, and Earth and space science, as well as 3 probes about the nature of science. The introductory chapter describes ways to use the probes and student work for professional learning. Probes from this book that can be used in grades K–2 include:

- "Is It a Solid?"
- "Does It Have a Life Cycle?"
- "Me and My Shadow"

Uncovering Student Ideas in Science, Volume 4 (Keeley and Tugel 2009)

The fourth book in the series contains 23 formative assessment probes in life, physical, and Earth and space science, as well as 2 probes that target the crosscutting concepts of models and systems. The introductory chapter describes the link between formative and summative assessment. Probes from this book that can be used with grades K–2 include:

- "Magnets in Water"
- "Moonlight"

Uncovering Student Ideas in Physical Science, Volume 1 (Keeley and Harrington 2010)

The fifth book in the series, and the first in a planned four-book series of physical science probes, contains 45 force and motion formative assessment probes. The introductory chapter describes why students struggle with force and motion ideas and the implications for instruction. Probes from this book that can be used in grades K–2 include:

- "How Far Did It Go?"
- "Rolling Marbles"
- "Talking About Forces"
- "Does It Have to Touch?"
- "Balance Beam"

Uncovering Student Ideas in Life Science, Volume 1 (Keeley 2011)

The sixth book in the series, and the first in a planned three-book series of life science probes, contains 25 life science formative assessment probes. The introductory chapter describes how formative assessment probes are used in a life science context. Probes from this book that can be used in grades K–2 include:

- "Cucumber Seeds"
- "No Animals Allowed"
- "Pumpkin Seeds"
- "Rocky Soil"
- "No More Plants"
- "Chrysalis"

Uncovering Student Ideas in Astronomy (Keeley and Sneider 2011)

The seventh book in the series contains 45 formative assessment probes for astronomy. The introductory chapter describes how formative assessment probes are used to understand students' mental models in astronomy. Probes from this book that can be used in grades K–2 include:

- "Where Do People Live?"
- "Sunrise to Sunset"
- "Seeing the Moon"
- "Sizing up the Moon"
- "Crescent Moon"

Features of This Book

This book contains 25 probes for the K–2 grade range organized in three sections: Section 1, Life Science (8 probes); Section 2, Physical Science (11 probes); and Section 3, Earth and Space Science (6 probes). The format is similar to the other seven volumes, with a few changes due to the focus on the primary grades. For example, the introduction focuses on young children's ideas in science and the use of "science talk." What follows are descriptions of the features of the Teacher Notes pages that accompany each probe in this book.

Purpose

This section describes the purpose of the probe—what you will find out about your students' ideas when you use the probe. It begins with the overall concept elicited by the probe, followed by the specific idea the probe targets. Before choosing a probe, it is important to be clear about what the probe is designed to reveal. Taking time to read the purpose will help you decide if the probe fits your intended learning target.

Related Concepts

Each probe is designed to target one or more concepts that are appropriate for grades K–2 students. A concept is a one-, two-, or three-word mental construct used to organize the related ideas addressed by the probe and the related national standards. These concepts are also included on the matrix charts that precede the probes on pages 2, 44, and 92.

Explanation

A brief scientific explanation accompanies each probe and provides clarification of the scientific content that underlies the probe. The explanations are designed to help the teacher understand what the "best" or most scientifically acceptable answers are, as well as clarify any misunderstandings about the content.

The explanations are not intended to provide detailed background knowledge about the content or designed to be shared with the student. The explanation is for the teacher. In writing these explanations, the author is careful not to make them too technical, as many primary-grade teachers have a limited content background in science. At the same time, the author takes great care not to oversimplify the science. The intent is to provide the information a science novice would need to understand the content related to the probe. If you have a need or desire for additional or deeper explanations of the content, refer to the NSTA resources listed for each section to build or enhance your content knowledge. For example, Bill Robertson's *Stop Faking It! Finally Understanding Science So You Can Teach It* books are excellent resources for furthering your own content understanding.

Curricular and Instructional Considerations for Grades K–2

Unlike the collections in the other books of this series, which address curricular and instructional considerations across the K–12 grade range, these probes are designed for the primary grades, and therefore the curricular and instructional considerations focus only on the K–2 grade span. (*Note:* Several probes are appropriate for the preK level as well.) This section provides a broad overview of the curricular emphasis in the primary science curriculum, the types of instructional experiences appropriate for K–2 students, and the difficulties to be aware of when teaching and learning the concepts related to the probe. This section will sometimes alert teachers to a K–2 idea from *A Framework for K–12 Science Education* (NRC 2012) that in the past has typically been addressed at the upper-elementary or middle-school level but has been moved down to the primary level. For example, static friction and waves are two concepts that typically have not been addressed

Preface

at the K–2 level but are included in the grades K–2 span in *A Framework for K–12 Science Education.* These are new considerations for curriculum and instruction that teachers of primary science will need to be aware of as they use the probes and design learning experiences for their students. In some cases, the developers of the *Next Generation Science Standards* (*NGSS*) decided to move disciplinary core ideas to a later grade span. This, too, is noted in the Teacher Notes.

Administering the Probe

Suggestions are provided for administering the probe to students, including response methods and ways to use props, demonstrate the probe scenario, make modifications for different learners, or use different formative assessment classroom techniques (FACTs) to gather the assessment data. This section also suggests referring to pages xxviii–xxxiii in the introduction for techniques that can be used to guide "science talk."

Related Ideas in the National Standards

This section lists the learning goals stated in two national documents. One has been used extensively to develop the learning goals in states' standards and curriculum materials: The revised, online version of *Benchmarks for Science Literacy* (AAAS 2009) includes a K–2 grade span. The *National Science Education Standards* (NRC 1996) were released after the original *Benchmarks* and overlap considerably in content. However, they are not included here, as the content is similar and the learning goals target too broad a grade span (K–4). Another referenced document in the teacher notes, *A Framework for K–12 Science Education* (NRC 2012), was used to identify the disciplinary core ideas that informed the development of the K–2 performance expectations in the *NGSS*. Since the *Benchmarks* is one of the primary documents on which most

state standards have been based prior to the release of the *NGSS*—which will be adopted by several states—it is still important to look at the related learning goals in this document. The third source of standards referenced is the *NGSS,* which were released shortly before the publication of this book. The Teacher Notes include the related performance expectation, which is the final assessment of student learning and is informed by the disciplinary core ideas listed under the *Framework.* The probes are not designed as summative assessments, so the listed related national standards are not intended to be considered alignments, but rather ideas that are related in some way to the probe. Some targeted probe ideas, such as the concept of living versus nonliving, are not explicitly stated as learning goals in the standards, yet they are important prerequisite ideas to understanding core ideas related to living things. In some cases, the *NGSS* performance expectation may not relate directly to the probe. Because performance expectations are designed for summative assessment purposes and not considered the curriculum or instruction, teachers need to provide experiences for students to learn the underlying ideas and concepts that will deepen the knowledge they will use to demonstrate the performance expectation. Formative assessment reveals the gaps that teachers can address in their instruction to move students toward the intended learning targets.

Related grade 3–5 learning goals, as well as some middle school learning goals, are also included in this section because it is useful to see the related idea that builds on the probe ideas at the next grade level. In other words, primary teachers can see how the foundation they are laying relates to a spiraling progression of ideas as students move from the primary grades to the intermediate elementary level. It may also be useful to teachers in grades 3–5 who may choose to use one of these probes to assess for gaps in students' understanding or misconceptions that are still tenaciously held. Sometimes the listed learning goals from upper grade spans appear to be unrelated to the probe or the K–2 learning goal. It is important to recognize that learning is a progression, and while it may seem unrelated, there is a connection. For example, when K–2 students learn about natural resources, it is in the context of objects and materials that are made from resources obtained from the Earth. The "Is a Brick a Rock?" probe is an example. However, in grade 4, the *NGSS* performance expectation shifts to energy sources and fuels obtained from the Earth. While the grade 4 *NGSS* performance expectation does not directly align with the probe, it is related to the bigger idea that humans use natural reources from the Earth.

Related Research

Each probe is informed by related research. Three comprehensive research summaries commonly available to educators—Chapter 15 in *Benchmarks for Science Literacy* (AAAS 1993), Rosalind Driver's *Making Sense of Secondary Science: Research Into Students' Ideas* (Driver, Squires, Rushworth, and Wood-Robinson 1994), and recent summaries in the *Atlas of Science Literacy, Volume 2* (AAAS 2007)—were drawn on for the research summaries. In addition, recent research is cited where available. Although many of the research citations describe studies that have been conducted in past decades and studied children not only in the United States but in other countries as well, most of the results of these studies are considered timeless and universal. Whether students develop their ideas in the United States or other countries, research indicates that many of these commonly held ideas are pervasive regardless of geographic boundaries and societal and cultural influences. Misconceptions held by students studied in past decades still

exist today. Even though your students may have had different experiences and contexts for learning, the descriptions from the research can help you better understand the intent of the probe and the kinds of thinking your students will likely reveal when they respond to the probe. As you use the probes, you are encouraged to seek new and additional published research or engage in your own action research to learn more about students' thinking and share your results with other teachers to extend and build on the research summaries in the teacher notes. To learn more about conducting action research using the probes, read the *Science and Children* article "Formative Assessment Probes: Teachers as Classroom Researchers" (Keeley 2011).

Suggestions for Instruction and Assessment

Uncovering and examining the ideas children bring to their learning is considered diagnostic assessment. Diagnostic assessment becomes formative assessment when the teacher uses the assessment data to make decisions about instruction that will move students toward the intended learning target. Therefore, for the probe to be considered a formative assessment probe, the teacher needs to think about how to best design, choose, or modify a lesson or activity to best address the preconceptions students bring to their learning or misconceptions that might surface or develop during the learning process. As you carefully listen to and analyze your students' responses, the most important next step is to decide on the instructional path that would work best in your particular context based on your students' thinking, the materials you have available, and the different types of learners you have in your classroom.

The suggestions provided in this section have been gathered from the wisdom of teachers, the knowledge base on effective science teaching, and research on specific strategies

Preface

used to address commonly held misconceptions. These are not lesson plans, but rather brief suggestions that may help you plan or modify your curriculum or instruction to help students learn ideas with which they may be struggling or that are incomplete. It may be as simple as realizing you need to provide an effective context, or there may be a specific strategy or activity that you could use with your students. Learning is a complex process, and most likely no single suggestion will help all students learn. But that is what formative assessment encourages: thinking carefully about the variety of instructional strategies and experiences needed to help students learn scientific ideas. As you become more familiar with the ideas your students have and the multi-faceted factors that may have contributed to their misunderstandings, you will identify additional strategies that you can use to teach for deeper understanding. In addition, this section also points out other probes in the *Uncovering Student Ideas in Science* series that can be modified or used as is to further address the concepts targeted by the probe.

Related NSTA Resources

This section provides a list of additional materials that can provide further information on content, curriculum, or instruction related to the probe. For example, Bill Robertson's *Stop Faking It!* series may be helpful in clarifying content with which teachers struggle; Karen Ansberry and Emily Morgan's *Picture-Perfect Science* series may suggest trade books and classroom activities that go together; and Dick Konicek's *Everyday Science Mysteries* series may provide an engaging story context to help students start investigating ideas. Articles from NSTA's *Science and Children* journal may suggest ways to provide primary-grade students the opportunity to learn the content targeted by the probe.

References

References are provided for the standards, cited research, and some of the instructional suggestions given in the Teacher Notes. You might also wish to read the full research summary or access a copy of the research paper or resource cited in the Related Research and Suggestions for Instruction and Assessment sections of the Teacher Notes.

Formative Assessment Probes in the Elementary Classroom

Formative assessment is an essential feature of a learning-focused elementary science environment. To help teachers learn more about using formative assessment probes with elementary students to inform instruction and promote learning, NSTA's elementary science journal, *Science and Children,* publishes my monthly column "Formative Assessment Probes: Promoting Learning Through Assessment." Your NSTA membership provides you with access to all of these journal articles, which have been archived electronically by NSTA. Go to the *Science and Children* page at *www.nsta. org/elementaryschool*. Scroll down to the journal archives and type "Formative Assessment Probes" in the keyword search box. This will pull up a listing of all of my columns. These articles can be saved in your library in the NSTA Learning Center or downloaded as a PDF. The following table lists the journal issue date, title of the column, and topic of the column for the articles that have been published to date. Check back regularly as more articles are added. These articles can also be used by preservice instructors, professional developers, and facilitators of professional learning communities to engage teachers in discussions about teaching and learning related to the probes and the content they teach.

Table 1. List of Articles in the Column "Formative Assessment Probes: Promoting Learning Through Assessment"

Date	Title	Topic
Sept. 2010	"Doing Science"	**Probe: "Doing Science"** "Scientific method": examine how misuse of the "scientific method" influences students' ideas related to the nature of science
Oct. 2010	"'More A-More B' Rule"	**Probe: "Floating Logs"** Floating and sinking: use of intuitive rules to reason about floating and sinking
Nov. 2010	"Does It Have a Life Cycle?"	**Probe: "Does It Have a Life Cycle?"** Life cycles: addressing the limitations of context in the curriculum
Dec. 2010	"To Hypothesize or Not"	**Probe: "Is It a Hypothesis?"** Hypothesis making: reveal misconceptions teachers have about the nature of science that can be passed on to students
Jan. 2011	"How Far Did It Go?"	**Probe: "How Far Did It Go?"** Linear measurement: difficulties students have with measurement particularly with a non-zero starting point
Feb. 2011	"Needs of Seeds"	**Probe: "Needs of Seeds"** Needs of living things: engaging in evidence-based argumentation
March 2011	"The Mitten Problem"	**Probe: "The Mitten Problem"** Energy transfer; insulators: teaching for conceptual change and how children's everyday experience affects their thinking
April 2011	"Is It Living?"	**Probe: "Is It Living?"** Characteristics of living things: examine ways to uncover hidden meanings students have for some words and concepts in science
July 2011	"With a Purpose"	**Probe: various examples** A variety of probes and concepts are used to show purposeful links to various stages in an assessment, instruction, and learning cycle
Sept. 2011	"Where Are the Stars?"	**Probe: "Emmy's Moon and Stars"** Solar system, relative distances: importance of examining students' explanations even when they choose the right answer; the impact representations have on children's thinking
Oct. 2011	"Pushes and Pulls"	**Probe: "Pushes and Pulls"** Forces: examining common preconceptions and use of language to describe forces and motion

Preface

Table 1 *(continued)*

Date	Title	Topic
Nov. 2011	"Teachers as Researchers"	**Probe: "Is It an Animal?"** Biological conception of an animal: explore how formative assessment probes can be used to engage in teacher action research
Dec. 2011	"Representing Microscopic Life"	**Probe: "Pond Life"** Single-celled organisms: use of representations to examine students' ideas
Jan. 2012	"Daytime Moon"	**Probe: "Objects in the Sky"** Objects in the sky: challenges the adage "seeing is believing" with "believing is seeing"; examines reasons why children hold on to their strongly held beliefs
Feb. 2012	"Can It Reflect Light?"	**Probe: "Can It Reflect Light?"** Reflection: addressing students' preconceptions with firsthand experiences that support conceptual change
April/May 2012	"Food for Plants: A Bridging Concept"	**Probe: "Food for Plants"** Food, photosynthesis, needs of plants: using bridging concepts to address gaps in learning goals, understanding students' common sense ideas
July 2012	"Where Did the Water Go?"	**Probe: "Where Did the Water Go?"** Water cycle: using the water cycle to show how a probe can be used to link a core content idea, scientific practice, and crosscutting concept
Sept. 2012	"Confronting Common Folklore: Catching a Cold"	**Probe: "Catching a Cold"** Infectious disease, personal health: using a probe to uncover common myths and folklore related to the common cold
Oct. 2012	"Talking About Shadows"	**Probe: "Me and My Shadow"** Sun-Earth system; talk moves: using a formative assessment probe to engage students in productive science talk
Nov. 2012	"Birthday Candles: Visually Representing Ideas"	**Probe: "Birthday Candles"** Light transmission; connection between light and vision: using drawings to support explanations
Dec. 2012	"Mountain Age: Creating Classroom Formative Assessment Profiles"	**Probe: "Mountain Age"** Weathering and erosion: organizing student data using a classroom profile for instructional decisions and professional development
Jan. 2013	"Solids and Holes: A P-E-O Probe"	**Probe: "Solids and Holes"** Floating/sinking; density: using the P-E-O technique to launch into inquiry
Feb. 2013	"Labeling Versus Explaining"	**Probe: "Chrysalis"** Life cycle of a butterfly: reveals how an overemphasis on labeling diagrams with correct terminology may mask conceptual misunderstandings related to the life cycle of a butterfly

Table 1 *(continued)*

Date	Title	Topic
March 2013	"When Equipment Gets in the Way"	**Probe: "Batteries, Bulb, and Wire"** Electrical circuits; lighting a bulb with a battery and a wire: how science kit materials may make it difficult for students to examine the way a complete circuit works
Summer 2013	"Is It a Solid? Claim Cards and Argumentation"	**Probe: "Is It a Solid?"** Matter, solids: using claims and evidence to engage in argumentation

Formative Assessment Reminder

Now that you have the background on this new series, the probes, and the Teacher Notes, we should not forget the formative purpose of these probes. Reminder that a probe is not formative unless you use the information from the probe to modify, adapt, or change your instruction so students have increased opportunities to learn the important scientific ideas necessary for building a strong foundation in the primary grades. As a companion to this book and all the other volumes, NSTA has copublished the book *Science Formative Assessment: 75 Practical Strategies for Linking Assessment, Instruction, and Learning* (Keeley 2008). In this book, you will find a variety of strategies to use along with the probes to facilitate elicitation, support metacognition, spark inquiry, encourage discussion, monitor progress toward conceptual change, encourage feedback, and promote self-assessment and reflection. In addition, be sure to read the suggestions in the introduction before using the probes. The introduction will help you learn more about ways to facilitate productive science discussions in the primary classroom and make links to the *Common Core State Standards* in English language arts. I hope the use of these K–2 probes and the techniques used along with them will stimulate new ways of assessing your students, create environments conducive to learning, and help you discover and use new knowledge about teaching and learning.

References

American Association for the Advancement of Science (AAAS). 1993. *Benchmarks for science literacy.* New York: Oxford University Press.

American Association for the Advancement of Science (AAAS). 2009. Benchmarks for science literacy online. *www.project2061.org/publications/bsl/online*

Driver, R., A. Squires, P. Rushworth, and V. Wood-Robinson. 1994. *Making sense of secondary science: Research into children's ideas.* London: RoutledgeFalmer.

Keeley, P. 2008. *Science formative assessment: 75 practical strategies for linking assessment, instruction, and learning.* Thousand Oaks, CA: Corwin Press and Arlington, VA: NSTA Press.

Keeley, P. 2011. *Uncovering student ideas in life science, vol. 1: 25 new formative assessment probes.* Arlington, VA: NSTA Press.

Keeley, P. 2011. Formative assessment probes: Teachers as classroom researchers. *Science and Children* 49 (3): 24–26.

Keeley, P., and R. Harrington. 2010. *Uncovering student ideas in physical science, vol. 1: 45 new force and motion assessment probes.* Arlington, VA: NSTA Press.

Keeley, P., and C. Sneider. 2012. *Uncovering student ideas in astronomy: 45 new formative assessment probes.* Arlington, VA: NSTA Press.

Keeley, P., F. Eberle, and C. Dorsey. 2008. *Uncovering student ideas in science, vol. 3: Another 25 formative assessment probes.* Arlington, VA: NSTA Press.

Preface

Keeley, P., F. Eberle, and L. Farrin. 2005. *Uncovering student ideas in science, vol. 1: 25 formative assessment probes.* Arlington, VA: NSTA Press.

Keeley, P., F. Eberle, and J. Tugel. 2007. *Uncovering student ideas in science, vol. 2: 25 more formative assessment probes.* Arlington, VA: NSTA Press.

Keeley, P., and J. Tugel. 2009. *Uncovering student ideas in science, vol. 4: 25 new formative assessment probes.* Arlington, VA: NSTA Press.

National Research Council (NRC). 1996. *National science education standards.* Washington, DC: National Academies Press.

National Research Council (NRC). 2012. *A framework for K–12 science education: Practices, crosscutting concepts, and core ideas.* Washington, DC: National Academies Press.

Acknowledgments

I would like to thank the teachers and science coordinators I have worked with for their willingness to field-test probes, provide feedback on the format and structure of these probes, share student data, and contribute ideas for assessment probe development. In particular, I would like to thank the teachers and administration at the J. Erik Jonsson Community School in Dallas, Texas, for their support, sharing of student data, and inspiring curriculum and teaching. I would like to thank all the elementary teachers who have agreed to collect artifacts and conduct interviews with students to use in professional development after this book is published. Thank you to the reviewers for providing useful feedback to improve the manuscript. I would especially like to thank Linda Olliver, the extraordinarily talented artist who creatively transforms my ideas into the visual representations seen on the student pages. And of course my deepest appreciation goes to all the dedicated staff at NSTA Press, who continue to support my formative assessment work, encourage me to continue with this series, and publish the best books in K–12 science education.

About the Author

Page Keeley recently retired from the Maine Mathematics and Science Alliance (MMSA), where she was the Senior Science Program Director for 16 years, directing projects and developing resources in the areas of leadership, professional development, linking standards and research on learning, formative assessment, and mentoring and coaching. She has been the PI and Project Director of three National Science Foundation–funded projects, including the Northern New England Co-Mentoring Network, PRISMS (Phenomena and Representations for Instruction of Science in Middle School), and Curriculum Topic Study: A Systematic Approach to Utilizing National Standards and Cognitive Research. In addition to NSF projects, she has directed state MSP projects, including TIES K–12: Teachers Integrating Engineering Into Science K–12 and a National Semi-Conductor Foundation grant, Linking Science, Inquiry, and Language Literacy (L-SILL). She also founded and directed the Maine Governor's Academy for Science and Mathematics Education Leadership, a replication of the National Academy for Science and Mathematics Education Leadership, of which she is a Cohort 1 Fellow.

Page is the author of 14 national best-selling books, including four books in the *Curriculum Topic Study* series, 8 volumes in the *Uncovering Student Ideas in Science* series, and both a science and mathematics version of *Formative Assessment: 75 Practical Strategies for Linking Assessment, Instruction, and Learning.* Currently, she provides consulting services to school districts and organizations throughout the United States on building teachers' and school districts' capacity to use diagnostic and formative assessment. She is a frequent invited speaker on formative assessment and teaching for conceptual change.

Page taught middle and high school science for 15 years before leaving the classroom in 1996. At that time, she was an active teacher leader at the state and national levels. She served two terms as president of the Maine Science Teachers Association and was a District II NSTA Director. She received the Presidential Award for Excellence in Secondary Science Teaching in 1992, the Milken National Distinguished Educator Award in 1993, the AT&T Maine Governor's Fellow in 1994, the National Staff Development Council's (now Learning Forward) Susan Loucks-Horsley Award for Leadership in Science and Mathematics Professional Development in 2009, and the National Science Education Leadership Association's (NSELA) Outstanding Leadership in Science Education Award in 2013. She has served as an adjunct instructor at the University of Maine, was a science literacy leader for the AAAS/Project 2061 Professional Development Program, serves on several national advisory boards, and is the Region A Director for NSELA. She is a science education delegation leader for the People to People Citizen Ambassador Professional Programs,

About the Author

leading trips to South Africa in 2009, China in 2010, India in 2011, and China again in 2013.

Prior to teaching, she was a research assistant in immunology at The Jackson Laboratory of Mammalian Genetics in Bar Harbor, Maine. She received her B.S. in life sciences from the University of New Hampshire and her master's degree in science education from the University of Maine. In 2008, Page was elected the 63rd president of the National Science Teachers Association (NSTA), the world's largest professional organization of science educators.

Introduction

"The having of wonderful ideas is what I consider the essence of intellectual development." —Eleanor Duckworth

K–2 Probes as Assessments for Learning

Imagine a first-grade classroom where Miss Ortega's students are sitting in a circle on the science rug to have a science talk about living things. Miss Ortega uses the pictures from the "Is It Living?" probe to have the children share their ideas about which things are living and which things are not living. As Miss Ortega shows the pictures and names the organisms or objects, she has each student turn to his or her partner to talk about ideas. They then discuss each picture as a whole class, with students sharing the rules they used to decide if something is living or nonliving.

At one point, the class is evenly divided about whether a seed is living or nonliving. Shalika argues, "The seed is not living because it can't move. The cat moves and it is alive, so I think living things have to move." Mac disagrees: "A tree is alive, but it can't move around like a cat. I think some things can be living and not move, so maybe a seed is alive." Oscar offers a new idea: "But seeds grow into plants." Miss Ortega asks Oscar to say more about that, and he adds a clarification: "Things that are living can grow. A seed grows into a plant, so that makes it living." Cora argues that some living things stop growing: "My dad is living, but he is done growing. I think you can stop growing and still be living."

Miss Ortega lists two ideas the children have proposed so far: moving and growing. She asks the class if there are other ways to decide if something is living. Albert offers a new idea: "Living things have to eat, so if it

eats, then it is alive." Kenny looks puzzled and asks a question to seek clarification: "But what about fire? It grows bigger when it eats wood." Rania responds, "Yeah, it moves around, too. Fire can move through a whole forest!"

The discussion continues for several minutes. The children are deeply involved in sharing their ideas, listening attentively to each other, seeking clarification from the teacher or other students when needed, constructing explanations to use in their arguments, and evaluating the ideas and arguments of others. Throughout the discussion, they are using and practicing speaking and listening skills.

Miss Ortega makes sure that all the children have an opportunity to make claims and express their thinking. Throughout the year, they have been working on claims and evidence during their science time. Her students know that a claim is the statement that answers the probe question, and to share their thinking, they must provide reasons, including evidence, for their claim. As the children are talking, Miss Ortega is carefully listening and making a list of the class's best ideas so far, which she will post on a chart for students to see and refer to while they visit the learning stations she will set up for the children to explore claims and ideas that support their claims. She notes the extent to which students are using scientific ideas or whether they are drawing on their own alternative conceptions or prior experiences.

By taking the time to find out what her students think about characteristics of living things, Miss Ortega collects valuable

Introduction

assessment data that she will use to plan lessons that will confront her students with their ideas and help them resolve some of the inconsistencies between their ideas and the scientific explanation she will guide them toward developing. By taking their ideas seriously and not correcting students' initial misconceptions, Miss Ortega is promoting learning by giving her students the opportunity to use scientific practices as they listen to each other's ideas, justify their own reasoning, and evaluate the validity of each other's arguments. This is the essence of formative assessment where good instruction and assessment are inextricably linked. Formative assessment is an approach to teaching in which students develop deeper conceptual understanding through the development of their own thinking and talking through their ideas, while simultaneously providing a window for the teacher to examine students' thinking and determine next steps based on where the learners are in their conceptual development.

Facilitating this type of approach to learning may sound demanding, and it seems it would be much simpler to just give students the information or engage in a hands-on activity. However, research shows that children learn best when they first surface their ideas before launching into investigations, activities, readings, and other opportunities to learn (Bransford, Brown, and Cocking 1999). Students have to do the thinking; the activity cannot do it for them, nor is the learning in the materials themselves. Surfacing initial ideas and recognizing when their ideas are changing as they construct new understandings is a powerful way to teach and learn. It is the explanation of the probe—not the answer selections the students choose from—that pro-

vides important assessment data and supports learning. One of the most effective ways to help students construct new understandings and simultaneously develop reasoning skills (that works particularly well with the formative assessment probes) is to provide children with the opportunity to interact in pairs, small groups, and as a whole class, where they listen to each other's ideas and have to justify their own. Communicating ideas in science is a central feature of using the formative assessment probes and one of the *Next Generation Science Standards (NGSS)* scientific practices. The probes not only provide insights for the teacher about students' understanding or misunderstanding of core ideas in science but also provide a treasure trove of data about students' use of scientific and engineering practices. For example, P-E-O (Predict-Explain-Observe) probes provide an opportunity for young children to make an initial claim (prediction), provide an explanation for their initial claim, test the claim, gather evidence (the data) from the observations, analyze the data to see if they support or refute the initial claim, and propose a new claim and explanation if the observations did not match the initial claim. See Table 2 for examples of ways the formative assessment probes support learning of the practices in the *NGSS*.

This book provides 25 highly engaging formative assessment probes that elicit preconceptions, support the development of young children's understanding of K–2 core disciplinary ideas, and encourage the use of scientific practices in the *NGSS*. However, before you skip ahead and use the probes, read through the rest of this introduction to learn more about children's ideas in science and science talk so that you can best use these probes to inform your teaching and support learning.

Table 2. Link Between K–2 Formative Assessment Probes and the *NGSS* Scientific and Engineering Practices

NGSS Scientific and Engineering Practice	K–2 Probes as Assessments for Learning	Examples
Practice 1: Asking questions and defining problems	The probe begins with an interesting question. Students ask additional questions to seek understanding and determine what they already know or need to know to make a claim and construct explanations to the probe.	**Probe 16, "Do the Waves Move the Boat?"** Students may ask questions of each other and the teacher about water waves. They use the probe question to further explore the science that can provide an explanatory answer to the probe.
Practice 2: Developing and using models	Some probes involve the use of models to develop explanations of the phenomenon. As the teacher listens to children's ideas, he or she is thinking about the best model to use to help them understand the probe phenomenon.	**Probe 24, "What Lights Up the Moon?"** As the teacher listens to students, he notices several students think there is a light glowing inside the moon. He thinks about how he can have the children model the reflection of sunlight using a white ball and a flashlight.
Practice 3: Planning and carrying out investigations	Some probes (P-E-O probes) can be used to launch an investigation and require children to think about how they can best make observations and collect data to test the claims they made in response to the probe.	**Probe 19, "Big and Small Magnets"** Students test their claims using a variety of big and small magnets. They decide how they will determine strength using paper clips, make observations, and record their data.
Practice 4: Analyzing and interpreting data	To support a claim with evidence when using a P-E-O probe, students collect, analyze, and interpret the data to derive meaning.	**Probe 18, "Rubber Band Box"** Students make rubber band box guitars like the one in the probe context and test their ideas about sound. They analyze their data to determine how pitch is related to the thickness of the rubber band.
Practice 5: Using mathematics and computational thinking	Students count and use numbers to find or describe patterns related to the probe. They also use measurement and measurement instruments such as thermometers, rulers, and weighing scales to gather data.	**Probe 12, "Snap Blocks"** Students count the individual blocks and make predictions about how the weight of the blocks snapped together compares to the total weight of the individual blocks weighed together. They try this several times using different numbers of snap blocks, weigh them on a scale, record their data, and notice that the pattern shows the weight is always the same.
Practice 6: Constructing explanations and designing solutions	Every probe requires students to construct an initial explanation (their personal theory) to support their claim (answer choice) and revise their explanations as they gather new evidence and information.	**Probe 10, "Watermelon and Grape"** The initial theory proposed by the class is that large things sink and small things float. After testing a variety of objects, students revise their initial explanation to explain that size alone does not determine whether an object floats or sinks.

Introduction

Table 2 *(continued)*

NGSS Scientific and Engineering Practice	K–2 Probes as Assessments for Learning	Examples
Practice 7: Engaging in argument from evidence	All the probes are used in a talk format that requires students to explain and defend their reasoning to others and promotes careful listening so that other students can build on a line of reasoning or offer alternative explanations. Together, the class searches for the best explanation from the comments and evidence offered during science talk.	**Probe 1, "Is It Living?"** As in the scenario at the beginning of this chapter, the teacher engages the students in science talk, explaining and defending their reasons for why something is or is not considered living. Later, after the teacher has provided opportunities for students to investigate their ideas, they will engage again in science talk, providing new evidence to support or revise their initial arguments.
Practice 8: Obtaining, evaluating, and communicating information	Following up the use of a probe often involves students seeking additional information that may come from trade books, videos, and other sources to provide information that supports their claims or provides new information to help them change their claim. Students must also be able to communicate information clearly to each other, which sometimes involves the use of drawings and other visual ways to share the information they obtain.	**Probe 6, "Do They Need Air?"** Many students claimed that animals that live in water do not need air, so the teacher obtained several trade books and video clips for children to learn about animals that live in water and how they meet their needs. She also brought in a goldfish that students could observe. Children revisit the probe and explain ways different animals get air and draw pictures of land and water animals, showing different structures they use to get air.

Young Children's Ideas

Children develop ideas about their natural world well before they are taught science in school. For example, many young children think that things like hats, coats, blankets, sweaters, or mittens warm us up by generating their own heat. This makes sense to children because when they put on a sweater or wrap themselves up in a blanket, they get warmer. They have not yet learned that heat moves from warmer to cooler and that materials such as a mitten can slow down the loss of body heat. In other instances, children are novice learners in science and have not yet gained enough background knowledge or been formally introduced to scientific principles to be able to explain a concept scientifically. For example, young children may think that only organisms with fur and four legs are considered animals because they do not yet have enough

knowledge about the scientific meaning of animal—such as having to acquire food from the environment—to recognize that organisms such as worms, insects, and even humans are considered animals in a scientific sense.

Some of the ideas young children have may be consistent or partially consistent with the science concepts that are taught. For example, they know when you drop an object, it falls to the ground. They are already developing ideas consistent with the idea of gravity. But often there is a significant gap between children's explanations for natural phenomena or concepts and the explanations that are developed through "school science." For example, failure to recognize that weight is conserved when a whole object is broken into individual pieces illustrates a significant gap between children's ideas about what happens to the total weight of an object when it is changed in some way and the understanding of conservation of matter with "parts and wholes" that is developed in the science class.

Many studies have been conducted of children's commonly held ideas about natural phenomena and science concepts. Most notable among the researchers of children's ideas is Dr. Rosalind Driver and her research group from the University of Leeds in England. This group has contributed extensively to our understanding of commonly held ideas children have about science that may affect their learning. The research into commonly held ideas, which is often referred to by practitioners as misconceptions, has enabled teachers to predict what their own students are likely to think about a phenomenon and how they might respond to an assessment that probes their thinking. The assessment probes in this book were developed from examining the research on children's ideas in science, particularly Driver's contributions (Driver, Squires, Rushworth, and Wood-Robinson 1994). As you use these probes, it is highly likely that your students' ideas will mirror the findings that are described in the Related Research Summaries part of the Teacher Notes that accompany each of the probes. While you may be surprised to find that your students hold many of these alternative ideas, it is important for you to realize that these are highly personal and make sense to the student. Merely correcting them does not make them go away. Students must have access to instructional experiences that will challenge their thinking and help them construct models and explanations that bridge the gap between their initial ideas and the scientific understandings that are achievable at their grade level.

Another important feature of children's ideas in science is that children learn best when knowledge is socially constructed. Much like the way science is done in the real world, children need opportunities to share their thinking, justify the reasons for their ideas, and listen to the ideas of others. Children also need to be aware of the range of ideas others have about the same phenomenon or concept and be able to evaluate them in light of their own ideas and the evidence presented. Scientific theories develop through interaction with other scientists; children's ideas develop through interaction with their classmates and the teacher. The probes provide ample opportunities for children to think through and talk about their ideas with others. Animated science talk and argumentation are the hallmarks of a formative assessment–centered learning environment in which the probes are effectively being used.

Children's Learning Experiences

Hands-on science has not always been minds-on science. The opportunity to ask questions, manipulate materials, and conduct investigations—a major emphasis of inquiry-based science—has not always resulted in deeper conceptual learning. That is because the learning is not in the materials or investigations themselves, but in the sense children make out of their experiences as they use the materials to perform investigations, make observations, and construct explanations. Perhaps the pendulum swung too far to the hands-on side in the last decade or so of elementary inquiry-based science. Inquiry without inquiry for conceptual change did little to help students give up their strongly held alternative ideas. One way to support inquiry for conceptual change is to start with uncovering children's ideas before launching into an investigation. To design probes for this book that could be used to support or enhance children's learning experiences, the following features for developing and using a probe were considered:

- Promoting curiosity and stimulating children's thinking
- Drawing out alternative ideas that could be investigated in the classroom

Introduction

- Linking to previous experiences in or out of school
- Using familiar objects, phenomena, and situations
- Reducing dependency on reading text
- Improving developmental appropriateness
- Relating to core ideas in the *NGSS* or the *Benchmarks for Science Literacy*
- Supporting the use of the scientific and engineering practices such as constructing explanations and engaging in argument from evidence
- Using models and representations to develop explanations
- Encouraging sense making and reflection on how ideas have changed

Effective teaching and learning do not just happen; they are carefully planned. Using the probes to inform the design of learning experiences for children involves the recognition that children already have ideas about the natural world that they bring with them to their learning experiences. This is significant when planning instructional experiences, particularly when combined with a constructivist view of learning in which the student must take an active role in constructing meaning for himself or herself. When children's existing ideas are acknowledged as you incorporate their ideas from the probes into the lessons, learning takes place as children change their ideas through experiences that allow them to test or discuss their ideas and support them with evidence, in much the same way that scientists develop theories. This may involve supporting an initial idea, modifying an idea, or rejecting an idea in favor of an alternative explanation. Whichever it is, the student needs to "own it," which means the reasoning must be done by the student (not the teacher, although he or she can guide it).

Formative Assessment Probes and Science Talk

When I first tried listening quietly and taking notes about what I heard students saying as they worked, my insight into their learning was phenomenal! I actually stopped talking and just listened. The data I collected showed some incorrect conceptions as well as understanding. It often opened windows into how a student had learned. The rich data I gathered helped me determine which next steps I needed to take to further learning. (Carlson, Humphrey, and Reinhardt 2003, p. 37)

This quote from an elementary teacher reveals the power of careful listening as students talk about their science ideas. The probes in this book differ from the collection of 215 other formative assessment probes in the *Uncovering Student Ideas* series because they are designed to be used in a talk format rather than having students write explanations to support their answer choice (however, they certainly can be combined with writing, especially with science notebooks). Even the formats used in this book highlight the importance of science talk. For example, you will see that several of the probes in this book use a cartoon format in which the characters share their ideas with each other and the student selects the character whose idea best matches his or her own (e.g., "Sink or Float?"). This format models what we want to encourage children to do: share their claims with one another and provide evidence that supports these claims. The author intentionally uses this format to help students recognize the importance of sharing ideas without passing judgment initially on whether they are right or wrong.

Whether you use the science workshop approach, science conferences, small-group discussions, pair talk, or other ways to involve students in science talk, the probes provide an interesting question that serves to elicit children's initial ideas and draw out their reasons for their thinking. Assessment data is gathered by carefully listening to students, and, when possible, audio-recording dialogue, taking notes, or transcribing parts of conversations as you listen will provide a treasure trove of data you can use to design instruction focused on where the learners are in their understanding.

The first step in using science talk for formative assessment is to ask questions that will capture students' interest, provoke thinking, and encourage explanations that will help you gain insight into their reasoning and understanding. Sometimes it can be challenging for teachers to know what type of question will elicit children's ideas that can provide rich information about a core concept they will be learning. This has been done for you in this book! Each probe is a question specifically designed to draw out children's ideas that will not only support their learning but also inform your teaching. Use the probe as your starting point for learning more about your students' ideas. New questions will spring from the probe and spark further conversation.

Sometimes a probe is used to develop an investigative question and launch into inquiry. Some probes provide an opportunity for students to make predictions and explain the reasons for their prediction before they make observations during their investigations. This type of probe is called a P-E-O (Predict-Explain-Observe) probe, and examples include, but are not limited to, "Snap Blocks" (p. 59), "Marble Roll" (p. 71), and "Seeds in a Bag" (p. 25) (Keeley 2008). P-E-O probes provide an opportunity for children to practice verbal communication to articulate their thoughts prior to the investigation and make

their thinking visible to others. In addition, it provides an opportunity for children to discuss the best way to test their ideas and then test them, supporting the *NGSS* scientific practice of designing and carrying out investigations.

Productive classroom talk using a formative assessment probe before launching into an investigation also has the benefit of leading to deeper engagement in the content before and during the investigation. As students collect, analyze, and share their data, they compare their findings with their initial claims and evidence and may become aware of the discrepancies between their own or others' ideas from the evidence gathered during the investigation. For example, the probe "When Is the Next Full Moon?" provides an opportunity for students to make and test a prediction about how long it takes to see a full Moon again (length of a lunar cycle). They examine reasons for their predictions before beginning an investigation that provides the data they need to understand the repeated pattern of Moon phases, a disciplinary core idea in the *NGSS* as well as a crosscutting concept of patterns and cycles. The probe can be revisited again after students have had an opportunity to make sense of their data and use it to explain the lunar cycle. By following the probe with a scientific investigation, the students have actual data from their investigation to construct a scientific explanation to support their new or revised claim. As they engage in talk and argument again with the same probe, the context of the probe—combined with the scientific knowledge they gained through their investigation—provide an opportunity for them to build stronger, evidence-based arguments.

Talk Moves

One of the best resources I recommend that every teacher of elementary science read and become familiar with is *Ready, Set, Science! Putting Research to Work in K–8 Classrooms*

Introduction

(Michaels, Shouse, and Schweingruber 2008). Since this book is published through our federal taxpayer–supported National Academies of Science, it is available for free as a PDF on the National Academies Press website (*www.nap.edu*), where you can download a copy of the book. It is also available for purchase through the NSTA bookstore. Chapter 5, "Making Thinking Visible: Talk and Argument," is an excellent read for you to deepen your understanding of the role of talk and argument in science as you use the probes in this book.

As students grapple with the ideas elicited by the probes in this book, the role of the teacher is to facilitate productive science talk in ways that will move students' thinking forward and help them clarify and expand on their reasoning. One of the ways to do this is through the use of "talk moves" (Keeley 2012). Table 3 shows six productive talk moves adapted from *Ready, Set, Science!* (Michaels, Shouse, and Schweingruber 2008) that can be used with the formative assessment probes in this book.

Table 3. Talk Moves and Examples

Talk Move (from *Ready, Set, Science!* [Michaels, Shouse, and Schweingruber 2008])	Example of Using the Talk Move with a Formative Assessment Probe
Revoicing	• "So let me see if I've got your thinking right. You're agreeing with Amy because _____?" • "Let me see if I understand. You are saying _____?"
Asking students to restate someone else's reasoning	• "Can you repeat in your own words what Latisha just said about why she agrees with Jamal?" • "Is that right, Latisha? Is that what you said?"
Asking students to apply their own reasoning to someone else's reasoning	• "Do you agree or disagree with Emma's reason for agreeing with Morrie, and why?" • "Can you tell us why you agree with what Sam said? What is your reasoning?"
Prompting students for further participation	• "Would someone like to add on to the reasons why some of you chose Fabian as the person you most agree with in the probe?" • "What about others—what would you like to add to these ideas so far?" • "What do others think about the ideas we have shared so far? Do you agree or disagree?"
Asking students to explicate their reasoning	• "Why do you agree with Penelope?" • "What evidence helped you choose Fabian as the person you most agree with in the probe?" • "Say more about that."
Using wait time	• "Take your time. We'll wait." • "I want everyone to think first, and then I will ask you to share your thinking."

Revoicing

Sometimes it is difficult to understand what the student is trying to say when they struggle to put their thoughts into words. If you, as the teacher, have difficulty understanding what the student is saying, then the students listening are apt to have even greater difficulty. Clarity in expressing ideas is often needed when encouraging young children to share their thinking. Therefore, this move not only helps the child clarify his or her thinking but also provides clarity for the listeners as well—both teacher and students. By revoicing the child's idea as a question, the teacher is giving the child more "think time" to clarify his or her ideas. It is also a strategy for making sure the student's idea is accessible to the other students who are listening and following the discussion.

Asking Students to Restate Someone Else's Reasoning

While the move above (revoicing) is used by the teacher, this move has the students reword or repeat what other students share during the probe discussion. It should then be followed up with the student whose reasoning was repeated or reworded. The benefit to using this talk move during discussions about the science probe ideas is that it gives the class more think time and opportunity to process each student's contribution to the science talk. It also provides another version of the explanation that may be an easier version for the children to understand. This talk move is especially useful with English language learners. As a formative assessment talk move, it provides the teacher with additional clarification of student ideas. Additionally, it acknowledges to the students that the teacher as well as the students in the class are listening to one another.

Asking Students to Apply Their Own Reasoning to Someone Else's Reasoning

The probes encourage students to make a claim and share their reasoning for their claim. This talk move is used with the probes to make sure students have had time to evaluate the claim based on the reasoning that was shared by a student. It helps students zero in and focus on the reasoning. Note that the teacher is not asking the other students whether they merely agree or disagree with someone's claim; they also have to explain why. This talk move helps students compare their thinking to someone else's and, in the process, helps them be more explicit in their own reasoning.

Prompting Students for Further Participation

After using revoicing to clarify the different ideas that emerge during discussion of the probe, the teacher prompts others in the class to contribute by agreeing, disagreeing, or adding on to what was already shared. This talk move encourages all students to evaluate the strength of each other's arguments. It promotes equitable and accountable discussion.

Asking Students to Explicate Their Reasoning

This talk move encourages students to go deeper with their reasoning and be more explicit in their explanations. It helps them focus on the evidence that best supports their claim and build on the reasoning of others.

Using Wait Time

This is actually a silent move, rather than a talk move. One of the hardest things for teachers to do is to refrain from not commenting immediately on children's responses. There are two types of wait time that should be used when engaging students in probe discussions. The first is for the teacher to wait at least five seconds after posing a question so the students

Introduction

have adequate think time. The second is for the teacher, as well as the students, to practice waiting at least five seconds before commenting on a students' response. This strategy is especially important to use with English language learners as well as students who may be shy or reluctant to contribute ideas in front of the whole class. By waiting, even though silence can be agonizing, the teacher supports students' thinking and reasoning by providing more time for them to construct an explanation or evaluate the arguments of others. This strategy provides greater inclusivity for all students in the class to participate in productive science talk by acknowledging the time they need to think through their ideas.

All of these talk moves can be used with the probes in various combinations to facilitate productive science talk in which all the students are accountable for each other's learning and the teacher is able to extract valuable formative assessment data to further plan instruction and support learning. However, to use these moves effectively, it is important to establish the conditions for a respectful learning environment. To do that, teachers should set group norms or ground rules for engaging in productive talk and equitable participation so that students will listen to and talk with one another respectfully and courteously as they use the probes. It is important for them to know that a scientific argument has a different meaning in science than in real life. In science, we argue to examine our ideas and seek understanding rather than argue to win with our point of view. Examples of norms you might establish in your classroom for science talk may include but are not limited to the following:

- Listen attentively as others talk.
- Make sure you can hear what others are saying.
- Speak so others can hear.
- Argue to learn, not to win.
- Criticize the reasoning, not the person.
- Make only respectful comments.

Communicating and listening to scientific ideas contribute to language development, an important goal of teaching in the primary grades, and is consistent with the *Common Core State Standards, ELA* (NGAC and CCSSO 2010). See Table 4 for examples of ways the formative assessment probes support the *Common Core* literacy standards for speaking and listening for primary students.

Formative assessment that supports productive scientific discussions takes time to develop and needs a lot of practice. As you incorporate these probes into your science lessons, I hope you will see the value in productive science talk that emanates from using these probes. By using these probes in talk formats with primary students, you are not only developing conceptual understanding of the life, physical, Earth, and space ideas for grades K–2 included in the *NGSS*, but also revealing and clarifying the ideas they bring to their learning, which you can use to improve and enhance your science teaching. Making students' ideas visible as you use these probes will help you build more effective lessons and support young students in using the scientific and engineering practices in sophisticated ways that show young learners are capable of far more than we often ask of them. In a nutshell, it's about teaching science well and giving your students the best possible start to be successful learners of science as they progress through school!

Table 4. Linking Formative Assessment Probes to the *Common Core* Speaking and Listening Standards

Common Core State Standards (Grades K, 1, and 2)	Formative Assessment Probes
◆ SL.K.1, 1.1, 2.1: Participate in collaborative conversations with diverse partners about kindergarten, grade 1, and grade 2 topics and texts with peers and adults in small and larger groups.	The probes are designed to be used in a talk format in small or large groups discussing ideas with the teacher and with each other about a science topic.
◆ SL.K.2: Confirm understanding of a text read aloud or information presented orally or through other media by asking and answering questions about key details and requesting clarification if something is not understood. ◆ SL.1.2: Ask and answer questions about key details in a text read aloud or information presented orally or through other media. ◆ SL.2.2: Recount or describe key ideas or details from a text read aloud or information presented orally or through other media.	As students talk about and share their ideas related to the probe, they ask questions about and discuss key details related to the probe context or answer choices. They may ask for clarification about the probe context or the answer choices or clarification of each other's explanations as they share their ideas through speaking and listening.
◆ SL.K.3: Ask and answer questions in order to seek help, get information, or clarify something that is not understood. ◆ SL.1.3: Ask and answer questions about what a speaker says in order to gather additional information or clarify something that is not understood. ◆ SL.2.3: Ask and answer questions about what a speaker says in order to clarify comprehension, gather additional information, or deepen understanding of a topic or issue.	Students ask questions about the probe task. They also ask questions of each other and seek clarification of explanations as they share their claims and provide their reasons for their claims. After students have had the opportunity to revisit the probe after the teacher has designed learning experiences, students ask and answer questions to deepen their understanding of the concepts elicited by the probe.
◆ SL.K.4: Describe familiar people, places, things, and events and, with prompting and support, provide additional detail. ◆ SL.1.4: Describe people, places, things, and events with relevant details, expressing ideas and feelings clearly. ◆ SL.2.4: Tell a story or recount an experience with appropriate facts and relevant, descriptive details, speaking audibly in coherent sentences.	The probes provide a context for discussing familiar phenomena, objects, and processes related to a science core idea. Students are encouraged to share their prior experiences connected to the probe and prompted by the teacher to provide details and further information.
◆ SL.K.5: Add drawings or other visual displays to descriptions as desired to provide additional detail. ◆ SL.1.5: Add drawings or other visual displays to descriptions when appropriate to clarify ideas, thoughts, and feelings. ◆ SL.2.5: Create audio recordings of stories or poems; add drawings or other visual displays to stories or recounts of experiences when appropriate to clarify ideas, thoughts, and feelings.	Students are encouraged to use drawings or other visual symbols, where appropriate, to support their ideas, clarify their responses, and communicate relevant details related to the probe.
◆ SL.K.6: Speak audibly and express thoughts, feelings, and ideas clearly. ◆ SL.1.6 and SL2.6: Produce complete sentences when appropriate to task and situation in order to provide requested detail or clarification.	Probes provide an engaging context for students to practice speaking clearly in complete sentences to support their ideas and emerging understanding of science.

Introduction

References

Bransford, J., A. Brown, and R. Cocking. 1999. *How people learn: Brain, mind, experience, and school.* Washington, DC: National Academies Press.

Carlson, M., G. Humphrey, and K. Reinhardt. 2003. *Weaving science inquiry and continuous assessment.* Thousand Oaks, CA: Corwin Press.

Driver, R., A. Squires, P. Rushworth, and V. Wood-Robinson. 1994. *Making sense of secondary science: Research into children's ideas.* London and New York: RoutledgeFalmer.

Duckworth, E. 2006. *"The having of wonderful ideas" and other essays on teaching and learning.* 3rd ed. New York: Columbia University Teachers College Press.

Keeley, P. 2008. *Science formative assessment: 75 practical strategies for linking assessment, instruction, and learning.* Thousand Oaks, CA: Corwin Press.

Keeley, P. 2012. Formative assessment probes: Talking about shadows. *Science and Children* 50 (2): 32–34.

Michaels, S., A. Shouse, and H. Schweingruber. 2008. *Ready, set, science: Putting research to work in K–8 classrooms.* Washington, DC: National Academies Press.

National Governors Association Center for Best Practices and Council of Chief State School Officers (NGAC and CCSSO). 2010. *Common core state standards.* Washington, DC: NGAC and CCSSO.

National Research Council (NRC). 2012. *A framework for K–12 science education: Practices, crosscutting concepts, and core ideas.* Washington, DC: National Academies Press.

Section 1

Life Science

Concept Matrix: Life Science
Probes #1–#8

RELATED CONCEPTS ↓	1. Is It Living?	2. Is It an Animal?	3. Is It a Plant?	4. Is It Made of Parts?	5. Seeds in a Bag	6. Do They Need Air?	7. Senses	8. Big and Small Seeds
animals		✓				✓		
breathing						✓		
characteristics of life	✓							
classification		✓	✓					
closed system					✓			
germination					✓			✓
information processing							✓	
living and nonliving things	✓							
needs of living things	✓				✓	✓		
parts and wholes				✓				
plants			✓					
plant tropisms							✓	
seeds					✓			✓
senses							✓	
structure				✓				
systems				✓				

Is It Living?

cat   seed   frog

fire   grass   river

rock   tree   cloud

What are you thinking?

Is It Living?

Teacher Notes

Purpose

The purpose of this assessment probe is to elicit children's ideas about living and nonliving things. The probe is designed to find out what characteristics children use to decide if something is living.

Related Concepts

characteristics of life, living and nonliving things, needs of living things

Explanation

There are five living things on this list: cat, frog, seed, grass, and tree. The fire, river, rock, and cloud are nonliving things. Living things can be defined by their structures and functions; all living things are made up of one or more cells. In addition, living things can carry out basic life processes such as obtaining or manufacturing food, extracting energy from food, growing, exchanging gases, reproducing, reacting to stimuli, moving, and eliminating waste. Not all living things show all of these characteristics all of the time. Some of these characteristics are easily observable, but others are not.

Curricular and Instructional Considerations for Grades K–2

At the K–2 level, the curricular emphasis should be on familiar plants and animals. Understanding what constitutes "living" involves identifying organisms' needs and observable characteristics. For example, young children can see that plants and animals are made up of different parts that allow them to meet their needs. Both plants and animals

need food, water, and air to live, and plants need sunlight as well. Plants and animals carry out basic observable functions such as using food and water, taking in air, growing, moving, responding, reproducing, and eliminating wastes. At the K–2 level, students are more likely to think of a developed organism as living rather than the initial stage of development, such as a seed, being a living thing. Therefore, students should have the opportunity to observe that a seed carries out functions of living things such as taking in water and growing. As students begin to recognize the diversity of life on our planet, it is important that they are able to distinguish between things that are living and things that are not living.

Administering the Probe

Review the things on the list with students to make sure they are familiar with each object listed. Name each object as you associate it with the picture. This practice is especially important for English language learners. Make sure the pictures are clear to the students—for example, some students may think the river is a road. If you have other pictures that illustrate each of the objects listed, you might show those to the students in addition to the ones on the student page. Instruct students to circle or color the things they think are living. Additionally, you may ask them to put an *X* over the ones they think are not living. Have students explain the rule they used to decide whether the things on the list are considered living or nonliving. Listen carefully for the criteria they use to decide if something is living.

See pages xxviii–xxxiii in the introduction for techniques used to guide "science talk" related to the probe.

Related Ideas in *Benchmarks for Science Literacy* (AAAS 2009)

. . . .

K–2 Cells

- Most living things need water, food, and air.

3–5 Cells

- Microscopes make it possible to see that living things are made mostly of cells.

Related Core Ideas in *A Framework for K–12 Science Education* (NRC 2012)

. . . .

K–2 LS1.A: Structure and Function

- All organisms have external parts. Different animals use their body parts in different ways to see, hear, grasp objects, protect themselves, move from place to place, and seek, find, and take in food, water, and air. Plants also have different parts (roots, stems, leaves, flowers, fruits) that help them survive, grow, and produce more plants.

K–2 LS1.B: Growth and Development of Organisms

- Plants and animals have predictable characteristics at different stages of development. Plants and animals grow and change. Adult plants and animals can have young.

3–5 LS1.A: Structure and Function

- Plants and animals have both internal and external structures that serve various functions in growth, survival, behavior, and reproduction.

3–5 LS1.B: Growth and Development of Organisms

Plants and animals have unique and diverse life cycles that include being born (sprouting in plants), growing, developing into adults, reproducing, and eventually dying.

Related *Next Generation Science Standards* (Achieve Inc. 2013)

. . . .

Kindergarten: From Molecules to Organisms: Structures and Processes

- K-LS1-1: Use observations to describe patterns of what plants and animals (including humans) need to survive.

Grade 1: From Molecules to Organisms: Structures and Processes

- 1-LS1-1: Use materials to design a solution to a human problem by mimicking how plants and/or animals use their external parts to help them survive, grow, and meet their needs.

Grade 2: Biological Evolution: Diversity and Unity

- 2-LS4-1: Make observations of plants and animals to compare the diversity of life in different habitats.

Grade 3: From Molecules to Organisms: Structures and Processes

- 3-LS1-1. Develop models to describe that organisms have unique and diverse life cycles but all have in common birth, growth, reproduction, and death.

Grade 4: From Molecules to Organisms: Structures and Processes

- 4-LS1-1: Construct an argument that plants and animals have internal and external

structures that function to support survival, growth, behavior, and reproduction.

Related Research

- Children have various ideas about what constitutes "living." Some may believe things that are active are alive—for example, fire, clouds, or the Sun. As children mature, they include eating, breathing, and reproducing as essential characteristics of living things. People of all ages use movement, and in particular movement as a response to a stimulus, as a defining characteristic of life. When doing so, people tend to omit plants from the "living" category. Some studies show that young children will infrequently cite growth as a criterion for life—the exception being when plants are identified as living (Driver, Squires, Rushworth, and Wood-Robinson 1994).

- A study by Stavy and Wax (1989) revealed that children seem to have different views for animal life and plant life. In general, animals were recognized more often than plants as being alive.

- Some studies indicate that the ability to reproduce is occasionally given by young children as a criterion for life. However, some nonliving things were said to be living because they "reproduced" (Driver, Squires, Rushworth, and Wood-Robinson 1994).

- Objects that children anthropomorphized are categorized as living things. For example, objects such as the Sun, cars, wind, and fire "felt" and "knew" things and were therefore alive. Studies indicate that there is a marked shift as students age from the view that things (including living things) carry out certain tasks "because they want to" to reasoning that "they need to in order to live" (Driver, Squires, Rushworth, and Wood-Robinson 1994).

- Carey (1985) suggested that progression in the concept of "living" is linked to growth

in children's ideas about biological processes. Young children have little knowledge of biology. In addition, it is not until around the age of 9 or 10 that children begin to understand death as the cessation of life processes.

- Piaget carried out some of the earliest studies on children's ideas about living. His results showed a predictable pattern in students' development of the concept of "living." From birth to age 5, students have almost no concept of living things; from ages 6 to 7, students believe things that are active or make noise are alive; from ages 8 to 9, students classify things that move as alive; from ages 9 to 11, students identify things that appear to move by themselves (including rivers and the Sun) as living; and past age 11, through adulthood, animals or animals and plants are considered living (Driver, Squires, Rushworth, and Wood-Robinson 1994).

Suggestions for Instruction and Assessment

- A particularly challenging example is the chrysalis stage of a butterfly. Many children see the butterfly as "dead" at this stage, even though they know it is part of the butterfly's life cycle. Combine this probe with "Chrysalis" from *Uncovering Student Ideas in Life Science* (Keeley 2011).

- There is also a K–8 version of this probe— "Is It Living?" in volume 1 of the *Uncovering Student Ideas in Science* series (Keeley, Eberle, and Farrin 2005)—that can be adapted for use with K–2 students.

- Add additional objects to the list to encourage further discussion and argumentation: Sun, mushroom, ant, feather, leaf, shell, car, wind, waves, and caterpillar. Include objects used in recent classroom investigations.

- Engage students in thought-provoking exercises that allow them to "discover"

why things are considered alive. Have children observe a number of objects that are classified as living—from whole organisms to parts of organisms, such as a carrot top placed in a dish of water. Have students identify characteristics that make these things "alive." Include items that may not be readily recognized as alive, such as plant seeds, flower bulbs, potatoes, mushrooms, and insect pupae.

- Be sure to distinguish needs from processes. For example, a seed may not need water for many years while it is dormant, but once environmental conditions are right and it can take in water, it will grow into a plant capable of sustaining life. Lessons should address the life processes (use of food for energy, reproduction, reaction to stimuli, breathing, movement, and waste elimination) in developmentally appropriate ways as students progress through the grades.

- Be aware of the tendency of younger children to anthropomorphize (attribute human form or personality to things that are not human). Explore the use of common phrases that imply nonliving things act in the same way as living things—for example, a fire "breathes" or waves "grow." Pay close attention to literature and images that make nonliving things seem living, such as putting a face on the Sun or clouds. Be a discerning user of children's literature.

- Have young students compare and contrast a stuffed animal toy or artificial plant with the real thing. Ask questions such as these: What can the living animal or plant do that the stuffed toy or artificial plant cannot? Why is one considered living and the other is not? Is the stuffed animal toy or artificial plant dead, or was it never alive? How do you know?

- The card-sort strategy can be used with this probe (Keeley 2008). Provide students with a set of cards that have pictures and names of living and nonliving objects on them. Have students work in small groups to sort the cards into three groups: things we think are living or were once living, things we think were never alive, and things about which we are unsure. A set of picture cards to choose from can be downloaded from the *Uncovering Student Ideas in Science* website (*www.uncoveringstudentideas.org/science_tools*).

Related NSTA Resources

Aram, R., and B. Bradshaw. 2001. How do children know what they know? *Science and Children* 39 (2): 28–33.

Keeley, P. 2011. Formative assessment probes: Is it living? *Science and Children* 48 (8): 24–26.

Konicek-Moran, R. 2008. *Everyday science mysteries: Stories for inquiry-based science teaching.* (See "Oatmeal Bugs," pp. 89–98.) Arlington, VA: NSTA Press.

Legaspi, B., and W. Straits. 2011. Living or nonliving? *Science and Children* 48 (8): 27–31.

References

Achieve Inc. 2013. *Next generation science standards.* www.nextgenscience.org/next-generation-science-standards.

American Association for the Advancement of Science (AAAS). 2009. Benchmarks for science literacy online. www.project2061.org/publications/bsl/online

Carey, S. 1985. *Conceptual change in childhood.* Cambridge, MA: MIT Press.

Driver, R., A. Squires, P. Rushworth, and V. Wood-Robinson. 1994. *Making sense of secondary science: Research into children's ideas.* London: Routledge.

Keeley, P. 2008. *Science formative assessment: 75 practical strategies for linking assessment, instruction, and learning.* Thousand Oaks, CA: Corwin Press and Arlington, VA: NSTA Press.

Keeley, P. 2011. *Uncovering student ideas in life science, vol. 1: 25 new formative assessment probes.* Arlington, VA: NSTA Press.

Keeley, P., F. Eberle, and L. Farrin. 2005. *Uncovering student ideas in science, vol. 1: 25 formative assessment probes.* Arlington, VA: NSTA Press.

National Research Council (NRC). 2012. *A framework for K–12 science education: Practices, cross-cutting concepts, and core ideas.* Washington, DC: National Academies Press.

Stavy, R., and N. Wax. 1989. Children's conception of plants as living things. *Human Development* 32: 88–89.

Is It an Animal?

mouse

fish

ant

puppy

flower

worm

person

snail

tree

What are you thinking?

Is It an Animal?

Teacher Notes

Purpose

The purpose of this assessment probe is to elicit children's ideas about animals. The probe is designed to find out what characteristics children use to determine whether an organism is an animal.

Related Concepts

animals, classification

Explanation

Everything is considered an animal except the flower and tree. Biologically, organisms are considered animals if they are heterotrophs (meaning they acquire food rather than manufacture it), are multicellular, pass through a blastula stage in their early embryonic development, and are mostly motile (or motile during some phase in their life cycle). Animals make up a vast diversity of species, ranging from microscopic animals (such as dust mites and tardigrades) to seemingly plantlike sponges and anemones to different types of vertebrates (including humans) to radially symmetrical starfish and sand dollars, cylindrical worms, and a huge variety of arthropods. While they may differ in size and internal and external features and behaviors, animals all share the common characteristics of being multicellular, lacking cell walls, acquiring food, having embryonic development, and possessing motility at some point in the life cycle. Although these criteria for determining whether an organism is an animal are intended for adult learners, at the K–2 level the criteria should focus on food acquisition. Animals must obtain their food from the environment by eating either plants or other animals.

Curricular and Instructional Considerations for Grades K–2

Primary-age children's familiarity with animals is derived from their firsthand experiences; books; or videos of pets, farm animals, zoos, and backyard creatures. By the time students come to school, they have already developed an operational definition of an animal. Their description of the characteristics that define an animal often includes fur, legs, ears and eyes, and movement. Most often their examples of animals are mammals and may include other vertebrates, as these are the animals they typically encounter through children's books, toys, and cartoons. Therefore, it is important for young children to learn about the vast diversity of animals so they are not limited by the context of their prior experiences and knowledge. There are several K–2 life science core ideas related to animals. At this level, distinguishing animals from other living things should be based on observable characteristics, including how animals get their food. Defining details about cells and embryological development are not appropriate at this grade level.

Administering the Probe

Review the things on the list with students to make sure they are familiar with each organism listed. Name each organism as you associate it with the picture. This practice is especially important for English language learners. Make sure the pictures are clear to the students. If you have other pictures that illustrate each of the organisms listed, you might show those to the students in addition to the ones on the student page. Instruct students to circle or color the things they think are animals. Additionally, you

may ask them to put an *X* over the ones they think are not animals. Have students explain the rule they used to decide whether the things on the list are considered animals. Listen carefully to the criteria they used to decide if something is an animal. See pages xxviii–xxxiii in the introduction for techniques used to guide "science talk" related to the probe.

Related Ideas in *Benchmarks for Science Literacy* (AAAS 2009)

K–2 Diversity of Life
- Some animals and plants are alike in the way they look and in the things they do, and others are very different from one another.

3–5 Diversity of Life
- A great variety of kinds of living things can be sorted into groups in many ways using various features to decide which things belong to which group.

Related Core Ideas in *A Framework for K–12 Science Education* (NRC 2012)

K–2 LS1.A: Structure and Function
- All organisms have external parts. Different animals use their body parts in different ways to see, hear, grasp objects, protect themselves, move from place to place, and seek, find, and take in food, water, and air.

3–5 LS1.A: Structure and Function
- Plants and animals have both internal and external structures that serve various functions in growth, survival, behavior, and reproduction.

K–2 LS4.D: Biodiversity and Humans
- There are many different kinds of living things in any area, and they exist in different places on land and in water.

3–5 LS4.D: Biodiversity and Humans
- Scientists have identified and classified many plants and animals.

Related *Next Generation Science Standards* (Achieve Inc. 2013)

Kindergarten: From Molecules to Organisms: Structures and Processes
- K-LS1-1: Use observations to describe patterns of what plants and animals (including humans) need to survive.

Grade 1: From Molecules to Organisms: Structures and Processes
- 1-LS1-1: Use materials to design a solution to a human problem by mimicking how plants and/or animals use their external parts to help them survive, grow, and meet their needs.

Grade 2: Biological Evolution: Diversity and Unity
- 2-LS4-1: Make observations of plants and animals to compare the diversity of life in different habitats.

Grade 3: From Molecules to Organisms: Structures and Processes
- 3-LS1-1: Develop models to describe that organisms have unique and diverse life cycles but all have in common birth, growth, reproduction, and death.

Grade 4: From Molecules to Organisms: Structures and Processes

- 4-LS1-1: Construct an argument that plants and animals have internal and external structures that function to support survival, growth, behavior, and reproduction.

Related Research

- People of all ages have a much narrower definition of an animal than biologists do. Students typically think of animals as terrestrial mammals. Students' ideas about animal qualities commonly include the following: having four legs, being large in size, having fur, making noise, and living on land (Bell 1981).

- Studies show that preservice elementary teachers, as well as experienced elementary teachers, also hold restricted meanings for the concept of animal. This may affect students' opportunities to learn the scientific concept of an animal (Driver, Squires, Rushworth, and Wood-Robinson 1994).

- Children often do not think of humans as animals; rather, they are contrasted with animals. Humans, insects, birds, and fish often are thought of as alternatives other than animals, not as subsets of animals (Driver, Squires, Rushworth, and Wood-Robinson 1994).

- Some research indicates that when a student is in second grade, there is a shift in his or her understanding of organisms, from representations based on perceptual and behavioral features to representations in which central principles of biological theory are most important (AAAS 2007, p. 30).

Suggestions for Instruction and Assessment

- Related probes about animals that can be used as is or adapted for K–2 include "Is It an Animal?" (Keeley, Eberle, and Farrin 2005) and "No Animals Allowed" (Keeley 2011).

- Primary-grade curricular units about animals are often focused on a specific animal or groups of animals (e.g., bears or farm animals). This may help students develop an understanding of the special characteristics of bears or farm animals that define them as animals, but students may fail to develop generalizable characteristics that apply to all animals. When using specific examples of animals, emphasize the major generalizable characteristics and point out examples of different types of animals that have these same characteristics.

- When students group organisms as animals and develop a rule that can be used to determine whether an organism is an animal, encourage them to use the rule to examine whether the characteristics they use are truly exclusive in distinguishing between animals and non-animals.

- Carefully examine children's trade books and literature to see if they portray a variety of vertebrate and invertebrate animals. Many children's animal books seem to focus more on vertebrates, particularly mammals.

- Explicitly develop the idea that all animals must obtain food from the environment and do so by eating plants or other animals. Continue to emphasize this idea from kindergarten through grade 2.

- Use the interview protocol developed by Charles Barman (Barman, Barman, Berglund, and Goldston 1999) to further examine students' ideas about animals.

- Be aware that accepting humans as animals may involve more than a shift from everyday to scientific thinking. Some students may have cultural or religious beliefs or traditions that may make them resistant to considering humans as animals. Teachers can respect these beliefs or traditions by balancing the scientific notion that

humans are biologically classified as animals with the notion that humans are a unique and special type of animal.

- Ask students to write down what something has to be like to be considered an animal, then have them draw an animal. Note how many of their drawings appear to match students' perceptions of animals as described in the research.

- Once students have developed an operational definition of an animal and compared it to a scientific definition appropriate for their grade level, they can be encouraged to further group animals into "types" using their own grouping schemes. They will see that animals are a diverse group of organisms and that some animals share special characteristics that other animals do not have (e.g., wings to fly, fur, backbones, eat plants) while noting characteristics that are common to all animals.

- The card-sort strategy can be used with this probe (Keeley 2008). Provide students with a set of cards that have pictures and names of animals on them. Have students work in small groups to sort the cards into three groups: things we think are animals, things we do not think are animals, and things about which we are unsure. A set of picture cards can be downloaded from the *Uncovering Student Ideas in Science* website (*www. uncoveringstudentideas.org/science_tools*).

- After students have completed this probe and had an opportunity to develop the concept of an animal, provide them with a list of new organisms. This can be used as an application or transfer of learning activity or a postassessment tool.

Related NSTA Resources

Barman, C., N. Barman, K. Berglund, and M. Goldston. 1999. Assessing students' ideas about animals. *Science and Children* 37 (1): 44–49.

Braude, S. 2007. The tree of animal life. *Science and Children* 45 (1): 42–48.

Keeley, P. 2011. Formative assessment probes: Teachers as classroom researchers. *Science and Children* 49 (3): 24–26.

Konicek-Moran, R. 2009. *More everyday science mysteries: Stories for inquiry-based science teaching.* (See "Worms Are for More Than Bait," pp. 91–100). Arlington, VA: NSTA Press.

Stovall, G., and C. Nesbit. 2003. Let's try action research. *Science and Children* 40 (5): 44–48.

References

Achieve Inc. 2013. *Next generation science standards. www.nextgenscience.org/next-generation-science-standards.*

American Association for the Advancement of Science (AAAS). 2007. *Atlas of science literacy.* Vol. 2. Washington, DC: AAAS.

American Association for the Advancement of Science (AAAS). 2009. Benchmarks for science literacy online. *www.project2061.org/publications/bsl/online*

Barman, C., N. Barman, K. Berglund, and M. Goldston. 1999. Assessing students' ideas about animals. *Science and Children* 37 (1): 44–49.

Bell, B. 1981. When is an animal not an animal? *Journal of Biological Education* 15 (3): 213–218.

Driver, R., A. Squires, P. Rushworth, and V. Wood-Robinson. 1994. *Making sense of secondary science: Research into children's ideas.* London: Routledge.

Keeley, P. 2008. *Science formative assessment: 75 practical strategies for linking assessment, instruction, and learning.* Thousand Oaks, CA: Corwin Press and Arlington, VA: NSTA Press.

Keeley, P. 2011. *Uncovering student ideas in life science, vol. 1: 25 new formative assessment probes.* Arlington, VA: NSTA Press.

Keeley, P., F. Eberle, and L. Farrin. 2005. *Uncovering student ideas in science, vol. 1: 25 formative assessment probes.* Arlington, VA: NSTA Press.

National Research Council (NRC). 2012. *A framework for K–12 science education: Practices, crosscutting concepts, and core ideas.* Washington, DC: National Academies Press.

Is It a Plant?

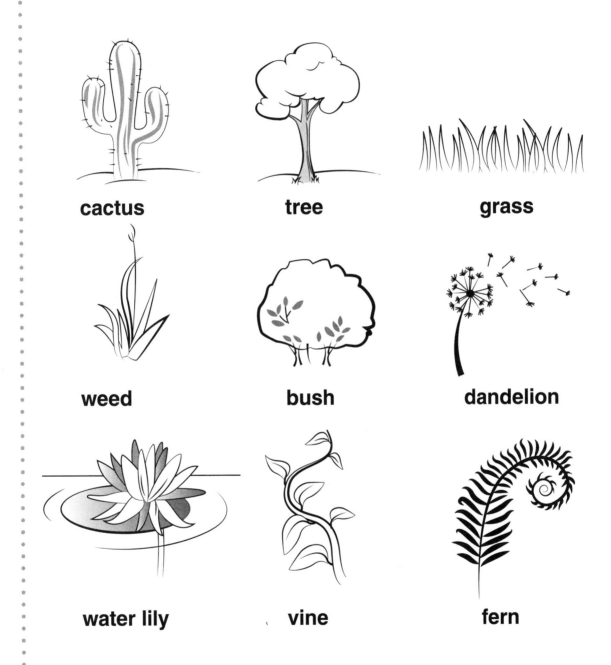

cactus

tree

grass

weed

bush

dandelion

water lily

vine

fern

What are you thinking?

Is It a Plant?

Teacher Notes

Purpose

The purpose of this assessment probe is to elicit children's ideas about plants. The probe is designed to find out what characteristics children use to decide if something is a plant.

Related Concepts

plants, classification

Explanation

All of the organisms listed are considered plants. They are all plants because they are multicellular, have cells that are surrounded by rigid cell walls, make their own food (sugar) through the process of photosynthesis, store food in the form of starch, and have limited motility. Plants can vary in size and shape, from tall trees to short grasses; from narrow, climbing vines to wide bushes; from seed producers to plants that produce spores such as ferns; and from plants that live in water to those that live on land. All plants have similar needs: sunlight, water, and air.

Curricular and Instructional Considerations for Grades K–2

At the primary-grade level, children distinguish plants from other organisms based on needs and observable characteristics such as the need for light; the existence of similar structures such as root, stems, leaves, or flowers; an inability or limited ability to move from place to place on their own; and the possession of green leaves and other bright pigments. At this level, details that define plants—such as their cell structure, photosynthesis, and reproduction and development—can wait until later grades for explanation. Although students may have opportunities to investigate plants in the primary grades and do encounter plants through their everyday interactions with the environment, they tend to have a narrow definition of a plant that often excludes trees, bushes, and weeds. Therefore, it is important to include a wide variety of plants in students' curricular materials and instructional activities.

Administering the Probe

Review the things on the list with students to make sure they are familiar with each organism listed. Name each plant as you associate it with the picture. This practice is especially important for English language learners. Make sure the pictures are clear to the students and that students have a sense of the plants' relative sizes, which the pictures do not convey. This is especially important in case students think plants are small and that larger, plantlike things are called trees or bushes. If you have other pictures that illustrate each of the organisms listed, you might show those to the students in addition to the ones on the student page. Instruct students to circle or color the things they think are plants. Additionally, you may ask them to put an *X* over the ones they think are not plants. Have students explain the rule they used to decide whether the things on the list are considered plants. Listen carefully to the criteria they used to decide if something is an animal. See pages xxviii–xxxiii in the introduction for techniques used to guide "science talk" related to the probe.

Related Ideas in *Benchmarks for Science Literacy* (AAAS 2009)

K–2 Diversity of Life

- Some animals and plants are alike in the way they look and in the things they do, and others are very different from one another.

3–5 Diversity of Life

- A great variety of kinds of living things can be sorted into groups in many ways using various features to decide which things belong to which group.

Related Core Ideas in *A Framework for K–12 Science Education* (NRC 2012)

K–2 LS1.A: Structure and Function

- Plants also have different parts (roots, stems, leaves, flowers, fruits) that help them survive, grow, and produce more plants.

3–5 LS1.A: Structure and Function

- Plants and animals have both internal and external structures that serve various functions in growth, survival, behavior, and reproduction.

K–2 LS4.D: Biodiversity and Humans

- There are many different kinds of living things in any area, and they exist in different places on land and in water.

3–5 LS4.D: Biodiversity and Humans

- Scientists have identified and classified many plants and animals.

Related *Next Generation Science Standards* (Achieve Inc. 2013)

Kindergarten: From Molecules to Organisms: Structures and Processes

- K-LS1-1: Use observations to describe patterns of what plants and animals (including humans) need to survive.

Grade 1: From Molecules to Organisms: Structures and Processes

- 1-LS1-1: Use materials to design a solution to a human problem by mimicking how plants and/or animals use their external parts to help them survive, grow, and meet their needs.

Grade 1: Ecosystems: Interactions, Energy, and Dynamics

- 2-LS2-1: Plan and conduct an investigation to determine if plants need sunlight and water to grow.

Grade 2: Biological Evolution: Diversity and Unity

- 2-LS4-1: Make observations of plants and animals to compare the diversity of life in different habitats.

Grade 3: From Molecules to Organisms: Structures and Processes

- 3-LS1-1: Develop models to describe that organisms have unique and diverse life cycles but all have in common birth, growth, reproduction, and death.

Grade 4: From Molecules to Organisms: Structures and Processes

- 4-LS1-1: Construct an argument that plants and animals have internal and external structures that function to support survival, growth, behavior, and reproduction.

Related Research

- Around second grade is when students begin to shift their thinking about organisms based on perceptual and behavioral features to more biological representations (AAAS 1993).
- Methods of classifying organisms vary by developmental level. For example, in upper elementary school, some students may group organisms such as plants by observable features, whereas other students may base their groupings on concepts (AAAS 1993).
- Elementary and middle school students hold a more restricted meaning than biologists for the word *plant*. Trees, vegetables, and grass are often not considered plants by students (AAAS 1993).
- In a study by Leach, Driver, Scott, and Wood-Robinson (1992), students used *plant*, *tree*, and *flower* as mutually exclusive groups. However, when students were given a restricted number of classification categories in a classification task, they assigned trees and flowers to the plant category (Driver, Squires, Rushworth, and Wood-Robinson 1994).
- In a study by Stead (1980), some children suggested that a plant is something that is cultivated; hence, grass and dandelion were considered weeds, not plants. Some children considered cabbage and carrots vegetables, not plants. They viewed vegetables as a comparable set rather than a subset of plants.

Suggestions for Instruction and Assessment

- A similar probe that can be modified for the K–2 grade level is "Is It a Plant?" in *Uncovering Student Ideas in Science, Volume 2: 25 More Formative Assessment Probes* (Keeley, Eberle, and Tugel 2007).

- Provide opportunities for children to describe and observe a variety of flowering and nonflowering plants (e.g., vegetables, flowers, ferns, mosses, trees, vines, weeds, bushes, grasses), not just the typical flowering plants (e.g., bean plants) that are commonly used in classroom investigations.
- The card-sort strategy can be used with this probe (Keeley 2008). Provide students with a set of cards that have pictures and names of plants on them. Have the students work in small groups to sort the cards into three groups: things we think are plants, things we do not think are plants, and things about which we are unsure.
- Develop the idea of "plant" as a broad category that can include a variety of groups (e.g., trees, vines, grasses, etc.). Let children develop the groupings. Provide students with picture cards of various types of plants, and have them sort the plants into groups.

Related NSTA Resources

Barman, C., M. Stein, N. Barman, and S. McNair. 2002. Assessing students' ideas about plants. *Science and Children* 40 (1): 46–51.

Barman, C., M. Stein, N. Barman, and S. McNair. 2003. Students' ideas about plants: Results from a national study. *Science and Children* 41 (1): 46–51.

Konicek-Moran, R. 2008. *Everyday science mysteries: Stories for inquiry-based science teaching.* (See "Trees From Helicopters," pp. 115–124.) Arlington, VA: NSTA Press.

Lawniczak, S., D. T. Gerber, and J. Beck. 1994. Plants on display. *Science and Children* 41 (9): 24–29.

References

Achieve Inc. 2013. *Next generation science standards.* www.nextgenscience.org/next-generation-science-standards.

American Association for the Advancement of Science (AAAS). 1993. *Benchmarks for science literacy.* New York: Oxford University Press.

American Association for the Advancement of Science (AAAS). 2009. Benchmarks for science literacy online. *www.project2061.org/publications/bsl/online*

Driver, R., A. Squires, P. Rushworth, and V. Wood-Robinson. 1994. *Making sense of secondary science: Research into children's ideas.* London: Routledge.

Keeley, P. 2008. *Science formative assessment: 75 practical strategies for linking assessment, instruction, and learning.* Thousand Oaks, CA: Corwin Press and Arlington, VA: NSTA Press.

Keeley, P., F. Eberle, and J. Tugel. 2007. *Uncovering student ideas in science, vol. 2: 25 more formative assessment probes.* Arlington, VA: NSTA Press.

Leach, J., R. Driver, P. Scott, and C. Wood-Robinson. 1992. *Progression in conceptual understanding of ecological concepts by pupils aged 5–16.* Leeds, England: University of Leeds, Centre for Studies in Science and Mathematics Education.

National Research Council (NRC). 2012. *A framework for K–12 science education: Practices, crosscutting concepts, and core ideas.* Washington, DC: National Academies Press.

Stead, B. 1980. *Plants.* LISP Working Paper 24. Hamilton, New Zealand: University of Waikato, Science Education Research Unit.

Is It Made of Parts?

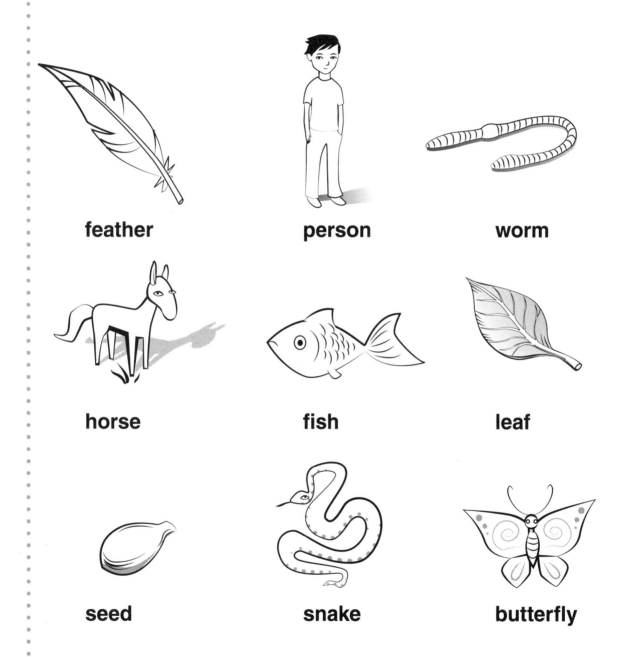

feather   person   worm

horse   fish   leaf

seed   snake   butterfly

What are you thinking?

Is It Made of Parts?

Teacher Notes

Purpose
The purpose of this assessment probe is to elicit children's developing ideas about structure in living systems. The probe is designed to find out if students recognize that living things are made up of parts.

Related Concepts
structure, parts and wholes, systems

Explanation
All of the things listed are made up of parts. An organism's body is made up of external and internal parts. These parts are the structure of an organism and work together as a system to enable an organism to carry out its life functions. Some external parts of organisms are easy to see. Others may be so small that magnification is needed. Some internal parts may not be obvious unless students look inside the object (e.g., a seed).

Curricular and Instructional Considerations for Grades K–2
Parts and wholes is a prerequisite concept for understanding structure and function of organisms and the crosscutting concept of systems (structure and function is also a crosscutting concept). In the primary grades, children learn that things are made of parts in both living and physical systems. Primary students should first observe external parts of organisms that they can easily see (e.g., ears of a dog), as well as parts that can be seen with magnifiers (eyes on an insect). Once students develop the idea that organisms have external parts, they may begin to explore familiar internal parts and how parts work together to allow an organism to live in its environment.

Administering the Probe
Review the things on the list with students to make sure they are familiar with each thing. Name each organism or part of an organism as you associate it with the picture. This practice is especially important for English language learners. Make sure the pictures are clear to the students and that students have a sense of the relative sizes, which the pictures do not convey. If you have other pictures that illustrate each of the things listed, you might show those to the students in addition to the ones on the student page. Instruct students to circle or color the things they think are made of parts. Additionally, you may ask them to put an X over the ones they think are not made of parts. Have students explain the rule and the criteria they used to decide whether the things on the list are made up of parts. See pages xxviii–xxxiii in the introduction for techniques used to guide "science talk" related to the probe.

Related Ideas in *Benchmarks for Science Literacy* (AAAS 2009)
. . . .

K–2 Cells
- Magnifiers help people see things they could not see without them.

K–2 Systems
- Most things are made of parts.

3–5 Systems

- In something that consists of many parts, the parts usually influence one another.

Related Core Ideas in *A Framework for K–12 Science Education* (NRC 2012)

K–2 LS1.A: Structure and Function

- All organisms have external parts. Different animals use their body parts in different ways to see, hear, grasp objects, protect themselves, move from place to place, and seek, find, and take in food, water, and air. Plants also have different parts (roots, stems, leaves, flowers, fruits) that help them survive, grow, and produce more plants.

3–5 LS1.A: Structure and Function

- Plants and animals have both internal and external structures that serve various functions in growth, survival, behavior, and reproduction.

Related *Next Generation Science Standards* (Achieve Inc. 2013)

Grade 1: From Molecules to Organisms: Structures and Processes

- 1-LS1-1: Use materials to design a solution to a human problem by mimicking how plants and/or animals use their external parts to help them survive, grow, and meet their needs.

Grade 4: From Molecules to Organisms: Structures and Processes

- 4-LS1-1: Construct an argument that plants and animals have internal and external structures that function to support survival, growth, behavior, and reproduction.

Related Research

- Children recognize that the body is made up of external parts before they appear to understand internal structures (Driver, Squires, Rushworth, and Wood-Robinson 1994).

Suggestions for Instruction and Assessment

- Start with familiar organisms (such as people) and have children identify different parts and their uses. Then proceed to different types of animals and plants.
- Have students go on a "parts and wholes" walk. Have them find things that are made of parts. Challenge them to find something that is not made of parts.
- Show students a plant and ask them if the plant is made up of parts. Have them explore different parts such as the roots, stems, leaves, flowers, fruits, and seeds. Then have them look for parts that make up the parts of a plant, developing the notion that things are made up of parts that may contain even smaller parts. However, be aware that labeling the parts of a plant does not ensure that students understand these are part of a larger system.
- Give students magnifiers and ask them to find parts of organisms (e.g., insects, worms, flowers, seeds) that they could not see with just their eyes to develop the notion that some parts are too small to see with just our eyes, and tools such as magnifiers help us see very tiny things.
- Project 2061's online Resources for Science Literacy has a good example of designing instruction for K–2 around the parts and wholes concept. You can access this lesson at *www.project2061.org/publications/rsl/ online/GUIDE/CH2/HLPPAR0.PDF.*
- This probe can be extended to a physical science context by providing students with a list of objects (e.g., toy, scissors, book)

and asking them to identify the things that are made of parts.

Related NSTA Resources

Ashbrook, P. 2008. The early years: Observing with magnifiers. *Science and Children* 45 (6): 18–20.

Konicek-Moran, R. 2009. *More everyday science mysteries: Stories for inquiry-based science teaching.* (See "Flowers: More Than Just Pretty," pp. 121–134.) Arlington, VA: NSTA Press.

Ritz, W. 2007. *A head start on science: Encouraging a sense of wonder.* Arlington, VA: NSTA Press.

References

Achieve Inc. 2013. *Next generation science standards. www.nextgenscience.org/next-generation-science-standards.*

American Association for the Advancement of Science (AAAS). 2009. Benchmarks for science literacy online. *www.project2061.org/publications/bsl/online*

Driver, R., A. Squires, P. Rushworth, and V. Wood-Robinson. 1994. *Making sense of secondary science: Research into children's ideas.* London: Routledge.

National Research Council (NRC). 2012. *A framework for K–12 science education: Practices, cross-cutting concepts, and core ideas.* Washington, DC: National Academies Press.

Seeds in a Bag

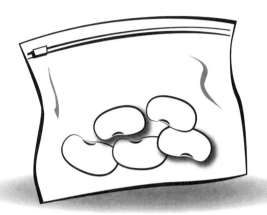

What will happen when the bag of seeds is put in soil?

All of the seeds will sprout.

Some of the seeds will sprout.

None of the seeds will sprout.

What are you thinking?

Seeds in a Bag

Teacher Notes

Purpose
The purpose of this assessment probe is to elicit children's ideas about the needs of seeds. The probe is designed to find out if students recognize that when seeds are planted in soil, they need to take in water from the soil in order to sprout. It also points out developing ideas students may have about open and closed systems.

Related Concepts
seeds, needs of living things, germination, closed system

Explanation
The best answer is "None of the seeds will sprout." The bag is a closed system that prevents the seeds from coming in direct contact with the water in the soil. Seeds encase a plant's embryo. Seeds need water, air, and the right temperature to sprout. Seeds need water (obtained from a moist environment) to sprout. Water is necessary for the metabolic reactions that result in the germination of the seed and its initial growth. Taking in water also leads to swelling inside the seed, which breaks the seed coat so that the seedling can emerge. Oxygen from the air is also needed for the cellular respiration that takes place in the cells of the plant embryo inside the seed and the emerging seedling. Because the bag is sealed, no water can get inside the bag and be taken in by the seeds. Students may think the seeds will sprout because they are planted in soil and/or need darkness and fail to recognize that the seeds need to be in direct contact with water.

Curricular and Instructional Considerations for Grades K–2
Investigating plant growth by observing and planting seeds under a variety of conditions is a common primary-level activity designed to help children understand that seeds need water, air, and the right temperature to sprout. Once they sprout, children learn that seedlings also need sunlight and air to grow. Children at this age also need opportunities to examine the conditions under which phenomena take place. The idea of an open or closed system—one of the crosscutting systems concepts that develop in sophistication from one grade level to the next—begins in early elementary grades by observing that some objects let things go in and out while others do not. Students need to recognize that the seeds must come in contact with water. Merely planting them in soil when they are in a sealed bag does not allow the water to reach the seed. This probe encourages students to think critically about the factors involved in germination.

Administering the Probe
Model this scenario for students by showing them some seeds in a sealed, zip-top bag. Make sure they understand the bag is sealed tight. Make sure they understand the inside of the bag is dry. Have students predict what they think will happen to the seeds if the bag is buried in soil and the soil is watered. Encourage students to explain the reasons for their prediction. Listen carefully to see if students recognize the bag as a closed system that prevents the seeds from coming in direct contact with

the water in the soil. Be aware that some students think the act of planting something in moist soil or watering the soil after the seeds are planted will sprout seeds without recognizing that the water must come in contact with the seeds. They may understand that seeds need water but not understand the seeds must take in the water. Refrain from giving students clues about the bag, as you will want them to think critically and discover the notion of the closed system not letting the water in after they have a chance to test their predictions. See pages xxviii–xxxiii in the introduction for techniques used to guide "science talk" related to the probe.

Related Ideas in *Benchmarks for Science Literacy* (AAAS 2009)

K–2 Flow of Matter and Energy
- Plants and animals both need to take in water, and animals need to take in food. In addition, plants need light.

Related Core Ideas in *A Framework for K–12 Science Education* (NRC 2012)

K–2 LS1.C: Organization for Matter and Energy Flow in Organisms
- Plants need water and light to live and grow.

3–5 LS1.C: Organization for Matter and Energy Flow in Organisms
- Animals and plants alike generally need to take in air and water, animals must take in food, and plants need light and minerals.

Related *Next Generation Science Standards* (Achieve Inc. 2013)

Kindergarten: From Molecules to Organisms: Structures and Processes
- K-LS1-1: Use observations to describe patterns of what plants and animals (including humans) need to survive.

Grade 1: Ecosystems: Interactions, Energy, and Dynamics
- 2-LS2-1: Plan and conduct an investigation to determine if plants need sunlight and water to grow.

Grade 4: From Molecules to Organisms: Structures and Processes
- 4-LS1-1: Construct an argument that plants and animals have internal and external structures that function to support survival, growth, behavior, and reproduction.

Related Research
- Some students fail to recognize a seed as a living thing; therefore, they do not recognize that seeds have needs similar to the needs of other living things (Driver, Squires, Rushworth, and Wood-Robinson 1994).
- Russell and Watt (1990) interviewed younger students about their ideas related to conditions for growth, focusing on germinations as well as vegetative growth. Of the 60 children interviewed, 90% identified water as necessary.
- A study found that students held strongly to the idea that light is always required by plants, even in the face of contrary evidence such as seedlings germinating in the dark (Roth, Smith, and Anderson 1983).
- Results from field-testing this probe indicate that many students think seeds will

sprout if planted in soil that has been watered. They fail to recognize that the seeds must come in contact with the water. The mere act of putting seeds in soil and adding water affected their thinking.

Suggestions for Instruction and Assessment

- A similar probe, "The Needs of Seeds" (Keeley, Eberle, and Tugel 2007), that addresses conditions necessary for germination can be adapted for this grade level.
- This probe can be used as a P-E-O (Predict-Explain-Observe) probe (Keeley 2008). Have students make their prediction, explain the reason for their prediction, then test the prediction by putting bean seeds in a sealed zip-top bag, covering the bag with soil, and watering the soil. Check the seeds several days later. Furthermore, compare seeds planted without being in a sealed bag with seeds in a sealed bag or seeds in an open bag. Conduct a class discussion to explain why the seeds in the open bag or no bag germinated and the seeds in the sealed bag did not.
- Start with the concept of a door to develop the idea of an open system. When the door is open, people in a room can leave and people outside a room can enter. Then develop the concept of a closed system. When the door is closed, people inside cannot leave the room and people outside the room cannot enter. After students have developed the idea of open and closed systems, ask them to think of how the seeds in a sealed bag are like the people in a room with a door. Ask them what things could get into or out of the bag when it is open versus closed.
- Show students what happens to a seed when it takes in water. Give students dry lima beans. Soak the lima beans overnight and compare the soaked lima beans to the dry lima beans. Discuss the evidence that the seed absorbed water. This can be further developed by weighing dry seeds and weighing them again after they have absorbed water.
- After students put the bag in the soil, they should water the soil. Challenge them to use the concept of open and closed systems to think about whether the seeds in the bag will absorb the water that is added to the soil. If students are reluctant to give up the idea that watering the soil will help the seeds sprout, confront them with a sealed bag of seeds submerged in water. Have them observe if the seeds come in contact with the water and how this might be similar to putting the sealed bag of seeds in wet soil.
- Some students may think the seeds did not germinate because they need light. Use caution when teaching the idea in the standards that plants need light. Once a seed sprouts to form a seedling, it needs light, but light is not necessary for germination. Students can further test this idea about seeds' need for light to sprout by placing seeds on a moist sponge or other substrate in the light versus in the dark.
- Provide students with opportunities to observe how water is necessary for seeds to sprout. Make observations of seeds with no water, little water, adequate water, and too much water to show that there is a certain amount of water that seeds need and that too much or too little water does not enable seeds to sprout.
- Challenge students to explain why seeds in seed packages do not sprout inside the package.

Related NSTA Resources

Ansberry, K., and E. Morgan. 2009. Teaching through trade books: Secrets of seeds. *Science and Children* 46 (6): 16–18.

Cavallo, A. 2005. Cycling through plants. *Science and Children* 42 (7): 22–27.

Keeley, P. 2011. Formative assessment probes: Needs of seeds. *Science and Children* 48 (6): 24–27.

Konicek-Moran, R. 2008. *Everyday science mysteries: Stories for inquiry-based science teaching.* (See "Seed Bargains," pp. 107–144). Arlington, VA: NSTA Press.

Ritz, W. 2007. *A head start on science: Encouraging a sense of wonder.* Arlington, VA: NSTA Press.

References

Achieve Inc. 2013. *Next generation science standards. www.nextgenscience.org/next-generation-science-standards.*

American Association for the Advancement of Science (AAAS). 2009. Benchmarks for science literacy online. *www.project2061.org/publications/bsl/online*

Driver, R., A. Squires, P. Rushworth, and V. Wood-Robinson. 1994. *Making sense of secondary science: Research into children's ideas.* London: Routledge.

Keeley, P. 2008. *Science formative assessment: 75 practical strategies for linking assessment, instruction, and learning.* Thousand Oaks, CA: Corwin Press and Arlington, VA: NSTA Press.

Keeley, P., F. Eberle, and J. Tugel. 2007. *Uncovering student ideas in science, vol. 2: 25 more formative assessment probes.* Arlington, VA: NSTA Press.

National Research Council (NRC). 2012. *A framework for K–12 science education: Practices, crosscutting concepts, and core ideas.* Washington, DC: National Academies Press.

Roth, K., E. Smith, and C. Anderson. 1983. *Students' conceptions of photosynthesis and food for plants.* East Lansing, MI: Michigan State University, Institute for Research on Teaching.

Russell, T., and D. Watt. 1990. *SPACE research report: Growth.* Liverpool, England: Liverpool University Press.

Do They Need Air?

What are you thinking?

Do They Need Air?

Teacher Notes

Purpose

The purpose of this assessment probe is to elicit children's ideas about needs of living things. The probe is designed to find out if students recognize that all animals need air to live, including animals that live under water.

Related Concepts

animals, needs of living things, breathing

Explanation

The best answer is Mei's: "I think all pond animals need air to live." All animals need air. When they take in air, they capture oxygen, which goes to the cells. Cells use the oxygen to help release energy from molecules that come from the food an animal ate. All animals get energy from food and therefore need oxygen for energy-releasing chemical reactions to take place in their cells. Land animals get their oxygen by breathing in air (some animals, such as amphibians, can absorb air through their skin). Aquatic animals must get their oxygen from the air that is dissolved in water. They use structures such as gills to extract dissolved oxygen from the water.

Curricular and Instructional Considerations for Grades K–2

In the elementary grades, students learn about the needs of animals. They learn animals need air. Understanding the need for air is a prerequisite to understanding that oxygen is the gas animals obtain from air. Students should learn the generalization that all animals need air, which means that even animals that live

under water need air to live. By connecting needs to structures designed to help organisms meet their needs, such as gills and lungs, K–2 students develop an understanding that even animals that live under water can get the air they need to live.

Administering the Probe

Point out each of the animals and their names to the students. Ask them which ones live by the water, which live under water, and which ones live both *under* water and *by* the water. Make sure students know that the tadpoles, water snail, and fish live under water. Do all the animals in and around the pond need air? Have children circle or color the person they agree with most and explain why they agree with that person. Explain to students that they should select the person whose idea most matches their thinking, not whose features they like most. Then begin a discussion of whether all the animals, or only some of them, need air to live. Note that this probe does not require students to know what oxygen is, but if it comes up during the class discussion, you may wish to clarify the idea that oxygen is found in the air. See pages xxviii–xxxiii in the introduction for techniques used to guide "science talk" related to the probe.

Related Ideas in *Benchmarks for Science Literacy* (AAAS 2009)

K–2 Cells

• Most living things need water, food, and air.

Related Core Ideas in *A Framework for K–12 Science Education* (NRC 2012)

K–2 LS1.A: Structure and Function

- Different animals use their body parts in different ways to see, hear, grasp objects, protect themselves, move from place to place, and seek, find, and take in food, water, and air.

3–5 LS1.A: Structure and Function

- Plants and animals have both internal and external structures that serve various functions in growth, survival, behavior, and reproduction

3–5 LS1.C: Organization for Matter and Energy Flow in Organisms

- Animals and plants alike generally need to take in air and water.

Related *Next Generation Science Standards* (Achieve Inc. 2013)

Kindergarten: From Molecules to Organisms: Structures and Processes

- K-LS1-1: Use observations to describe patterns of what plants and animals (including humans) need to survive.

Grade 1: From Molecules to Organisms: Structures and Processes

- 1-LS1-1: Use materials to design a solution to a human problem by mimicking how plants and/or animals use their external parts to help them survive, grow, and meet their needs.

Grade 4: From Molecules to Organisms: Structures and Processes

- 4-LS1-1: Construct an argument that plants and animals have internal and external structures that function to support survival, growth, behavior, and reproduction.

Related Research

- Although young children know that air is necessary for life, studies show that they appear to have a limited idea of what happens to inhaled air, often thinking it remains in the head (Driver, Squires, Rushworth, and Wood-Robinson 1994).
- A study of Nigerian students found that children learned from an early age that they breathe oxygen and that oxygen was often equated with air (Adeniyi 1985).
- In a classroom activity with crayfish, students did not identify air as a necessary part of the crayfish habitat (Endreny 2006).

Suggestions for Instruction and Assessment

- Connect the need all animals have for air to the idea that animals obtain their needs from the environment in different ways. For example, fish live in water and therefore need a way to get the air they need to live from the water. They have gills that are used to get air from the water. Some animals live part of the time in water and part of the time on land (amphibians). The structures they use to get air depend on whether they live in water (tadpole) or can live out of water (frog).
- Use picture books or videos to have students explore the variety of ways animals breathe in air to get oxygen. As students come across each animal, help them identify how the animal breathes: gills, a nose, a blow hole, a beak, or even by absorbing air through their skin. Explain how some animals can get the air they need under

water and some can breathe the air on land just like we do.

- To help students understand that aquatic animals also "breathe," show them the bubbles that are visible when released by some aquatic animals. Guide students toward realizing the animals take in air and "exhale" a gas like we do when we breathe. However, they take oxygen from the air that is dissolved in the water that is taken in through their mouth, passed over the gills to extract oxygen, then released as another gas (carbon dioxide), which is in the bubbles.

Related NSTA Resources

Alexander, D. 2010. *Hop into action: The amphibian curriculum guide for grades K–4.* Arlington, VA: NSTA Press.

Endreny, A. 2006. Crazy about crayfish. *Science and Children* 43 (7): 32–35.

Rich, S. 2012. *Bringing outdoor science in: Thrifty classroom lessons.* (See Chapter 5, "Animals That Live in Water.") Arlington, VA: NSTA Press.

References

Achieve Inc. 2013. *Next generation science standards. www.nextgenscience.org/next-generation-science-standards.*

Adeniyi, E. 1985. Misconceptions of selected ecological concepts held by some Nigerian students. *Journal of Biological Education* 19 (4): 311–316.

American Association for the Advancement of Science (AAAS). 2009. Benchmarks for science literacy online. *www.project2061.org/publications/bsl/online*

Driver, R., A. Squires, P. Rushworth, and V. Wood-Robinson. 1994. *Making sense of secondary science: Research into children's ideas.* London: Routledge.

Endreny, A. 2006. Crazy about crayfish. *Science and Children* 43 (7): 32–35.

National Research Council (NRC). 2012. *A framework for K–12 science education: Practices, crosscutting concepts, and core ideas.* Washington, DC: National Academies Press.

Senses

What are you thinking?

Senses

Teacher Notes

Purpose

The purpose of this assessment probe is to elicit children's ideas about sensory responses. The probe is designed to reveal whether students recognize that plants as well as animals can sense information from their environment and respond.

Related Concepts

senses, information processing, plant tropisms

Explanation

Pia has the best response: "I think animals and plants both sense things." Animals have internal and external sensory receptors that help them detect information from their environment. Senses such as taste, smell, sight, hearing, and touch enable animals to process, store, or react to information. Plants may not have well-developed nervous systems like animals, but they are capable of sensing external inputs from their environment such as light, direction, touch, and moisture. Plants exhibit a variety of *tropisms*, which are movement reactions to sensory stimuli, such as growing toward light (phototropism), roots growing downward (gravitropism), hydrotropism (movement toward water), growing toward or away from certain chemicals (chemotropism), and thigmotropism (response to touch).

Curricular and Instructional Considerations for Grades K–2

In the primary grades, children learn about parts of the body that get information from the environment, such as eyes for seeing and ears for hearing. Humans and animals respond to this information with behaviors that help them survive. Children at this age should also have opportunities to learn that plants can also sense their environment and react in ways that help them survive. For example, children should have opportunities to observe how plants grow toward the light or that roots always grow downward.

Administering the Probe

Have children circle or color the person with whom they agree the most and explain why they agree with that person. Explain to students that they should select the person whose idea most matches their thinking, not whose features they like most. Make sure students understand what it means to "sense things." To activate their thinking in relation to this probe, have students describe their own senses and their uses. Then encourage them to think about whether animals and plants have senses that help them live in their environment. See pages xxviii–xxxiii in the introduction for techniques used to guide "science talk" related to the probe.

Related Ideas in *Benchmarks for Science Literacy* (AAAS 2009)

K–2 Basic Functions

- The human body has parts that help it seek, find, and take in food when it feels hunger—eyes and a nose for detecting food, legs to get to it, arms to carry it away, and a mouth to eat it.
- Senses can warn individuals about danger; muscles help them to fight, hide, or get out of danger.

- The brain enables human beings to think and sends messages to other body parts to help them work properly.

Related Core Ideas in *A Framework for K–12 Science Education* (NRC 2012)

K–2 LS1.D: Information Processing

- Animals have body parts that capture and convey different kinds of information needed for growth and survival—for example, eyes for light, ears for sounds, and skin for temperature or touch. Animals respond to these inputs with behaviors that help them survive (e.g., find food, run from a predator). Plants also respond to some external inputs (e.g., turn leaves toward the Sun).

3–5 LS1.D: Information Processing

- Different sense receptors are specialized for particular kinds of information, which may then be processed and integrated by an animal's brain.

K–2 PS4.C: Information Technologies and Instrumentation

- People use their senses to learn about the world around them. Their eyes detect light, their ears detect sound, and they can feel vibrations by touch.

Related *Next Generation Science Standards* (Achieve Inc. 2013)

Grade 1: From Molecules to Organisms: Structures and Processes

- 1-LS1-1: Use materials to design a solution to a human problem by mimicking how plants and/or animals use their external

parts to help them survive, grow, and meet their needs.

Grade 4: From Molecules to Organisms: Structures and Processes

- 4-LS1-1: Construct an argument that plants and animals have internal and external structures that function to support survival, growth, behavior, and reproduction.
- 4-LS1-2: Use a model to describe that animals' receive different types of information through their senses, process the information in their brain, and respond to the information in different ways.

Related Research

There is scant research on children's ideas related to animal and plant sensory communication and behavior. However, studies of children's ideas about living reveal a lack of understanding of plants' behavioral responses. Because some children use movement and behavioral reaction as a criteria for living, they do not recognize that plants have behavioral reactions and thus do not consider them as living things (Driver, Squires, Rushworth, and Wood-Robinson 1994).

Suggestions for Instruction and Assessment

- There are three related probes on plant tropisms from other books in the *Uncovering Student Ideas* series that can be adapted for this grade level: "Plants in Light and Dark" (Keeley, Eberle, and Tugel 2007), "Pumpkin Seeds" (Keeley 2011), and "Rocky Soil" (Keeley 2011).
- Typically, when primary-age children learn about the five senses, the focus is on humans. Make sure children learn that all animals use their senses to obtain information from the environment and that plants also have similar ways of obtaining information from the environment.

- Provide opportunities for students to investigate plant "senses" that result in responses such as growing toward the light, roots growing downward and around hard objects, and tendrils that wrap around objects with which they come in contact. An especially interesting houseplant children can observe is the "sensitive plant," a type of mimosa that folds up its leaves when touched.

Related NSTA Resources

McWilliams, S. 2003. Journey into the five senses. *Science and Children* 40 (5): 38–43.

Ritz, W. 2007. *A head start on science: Encouraging a sense of wonder.* (See Section 1, "The Senses.") Arlington, VA: NSTA Press.

Tolman, M., and G. Hardy. 1999. Teaching tropisms. *Science and Children* 37 (3): 14–17.

References

Achieve Inc. 2013. *Next generation science standards. www.nextgenscience.org/next-generation-science-standards.*

Driver, R., A. Squires, P. Rushworth, and V. Wood-Robinson. 1994. *Making sense of secondary science: Research into children's ideas.* London: Routledge.

Keeley, P. 2011. *Uncovering student ideas in life science, vol. 1: 25 new formative assessment probes.* Arlington, VA: NSTA Press.

Keeley, P., F. Eberle, and J. Tugel. 2007. *Uncovering student ideas in science, vol. 2: 25 more formative assessment probes.* Arlington, VA: NSTA Press.

National Research Council (NRC). 2012. *A framework for K–12 science education: Practices, crosscutting concepts, and core ideas.* Washington, DC: National Academies Press.

Big and Small Seeds

What are you thinking?

Big and Small Seeds

Teacher Notes

Purpose

The purpose of this assessment probe is to elicit children's ideas about seed germination. The probe is designed to find out if children think the size of a seed determines the time it takes to germinate and grow.

Related Concepts

seeds, germination

Explanation

The best answer is Ace's: "I think seed size doesn't matter." Germination, or sprouting, is the process by which a young plant emerges from the seed and begins growth as a seedling. Germination generally depends on conditions such as amount of water, oxygen, and temperature. Some seeds require other conditions, such as passing through an animal's digestive tract (strawberry seeds), a period of cold (some oak acorns), or soaking in water for a long period of time (coconuts). Generally, germination time cannot be predicted by the size of a seed. For example, a large pumpkin seed can take 5 to 7 days to sprout into a seedling, and a tiny carrot seed can take 12 to 15 days. Conversely, a tiny mustard seed can take 4 to 5 days to germinate, and a larger pea seed can take 7 to 14 days.

Curricular and Instructional Considerations for Grades K–2

Primary students are encouraged to make observations of a variety of objects, materials, and living things. By observing seeds, they learn to describe characteristics such as size, shape, color, and texture. Planting seeds or germinating seeds in soil-less germination containers are common inquiry-based activities that help children learn about the needs of a seed and the life cycle of a plant. Collecting germination data on a variety of seeds of different sizes and analyzing the data to see if size makes a difference supports the scientific practices of planning and carrying out investigations and analyzing and interpreting data.

Administering the Probe

If possible, first show children a variety of seeds ranging from very small seeds to large seeds, such as a coconut. Ask them if the time it takes a seed to sprout depends on the size of the seed, then ask them to circle or color in the person in the probe with whom they agree the most and explain why they agree with that person. Explain to students that they should select the person whose idea best matches their thinking, not whose features they like most. See pages xxviii–xxxiii in the introduction for techniques used to guide "science talk" related to the probe.

Related Ideas in *Benchmarks for Science Literacy* (AAAS 2009)

K–2 Diversity of Life

- Some animals and plants are alike in the way they look and in the things they do, and others are very different from one another.

Related Core Ideas in *A Framework for K–12 Science Education* (NRC 2012)

K–2 LS1.B: Growth and Development of Organisms

- Plants and animals have predictable characteristics at different stages of development.

3–5 LS1.B: Growth and Development of Organisms

- Plants and animals have unique and diverse life cycles that include being born (sprouting in plants), growing, developing into adults, reproducing, and eventually dying.

Related *Next Generation Science Standards* (Achieve Inc. 2013)

Grade 2: Biological Evolution: Diversity and Unity

- 2-LS4-1: Make observations of plants and animals to compare the diversity of life in different habitats.

Grade 3: From Molecules to Organisms: Structures and Processes

- 3-LS1-1: Develop models to describe that organisms have unique and diverse life cycles but all have in common birth, growth, reproduction, and death.

Related Research

Some students use the intuitive idea identified by Stavy and Tirosh (2000), "More A-More B," to reason that larger seeds (More A) take more time to germinate (More B).

Suggestions for Instruction and Assessment

- This probe can be combined with other elementary probes about seed germination and growth, such as "Pumpkin Seeds" (Keeley 2011), "Cucumber Seeds" (Keeley 2011) and "Needs of Seeds" (Keeley, Eberle, and Tugel 2007).
- Provide students with a variety of seeds of different sizes and have them arrange them from smallest to largest to encourage observations related to the size of seeds.
- This probe can be used with the P-E-O strategy (Keeley 2008). Have students commit to a prediction that matches the probe, explain the reason for their prediction, then design and carry out an investigation to test the prediction. Each group can be assigned a different seed ranging from small to large seeds. After students have made their observations and collected and analyzed their germination data, give them an opportunity to revisit the probe by stating a claim and providing the evidence to support the claim.
- Provide old seed catalogs as a source of data that show different germination times for seeds.

Related NSTA Resources

Ansberry, K., and E. Morgan. 2009. Teaching through trade books: Secrets of seeds. *Science and Children* 46 (6): 16–18.

Keeley, P. 2011. Formative assessment probes: Needs of seeds. *Science and Children* 48 (6): 24–27.

Konicek-Moran, R. 2008. *Everyday science mysteries: Stories for inquiry-based science teaching.* (See "Seed Bargains," pp. 107–114). Arlington, VA: NSTA Press.

Ritz, W. 2007. *A head start on science: Encouraging a sense of wonder.* Arlington, VA: NSTA Press.

References

Achieve Inc. 2013. *Next generation science standards.* *www.nextgenscience.org/next-generation-science-standards.*

American Association for the Advancement of Science (AAAS). 2009. Benchmarks for science literacy online. *www.project2061.org/publications/bsl/online*

Keeley, P. 2008. *Science formative assessment: 75 practical strategies for linking assessment, instruction, and learning.* Thousand Oaks, CA: Corwin Press and Arlington, VA: NSTA Press.

Keeley, P. 2011. *Uncovering student ideas in life science, vol. 1: 25 new formative assessment probes.* Arlington, VA: NSTA Press.

Keeley, P., F. Eberle, and J. Tugel. 2007. *Uncovering student ideas in science, vol. 2: 25 more formative assessment probes.* Arlington, VA: NSTA Press.

National Research Council (NRC). 2012. *A framework for K–12 science education: Practices, crosscutting concepts, and core ideas.* Washington, DC: National Academies Press.

Stavy, R., and D. Tirosh. 2000. *How students (mis)understand science and mathematics: Intuitive rules.* New York: Teachers College Press.

Section 2

Physical Science

Concept Matrix: Physical Science
Probes #9–#19

RELATED CONCEPTS ↓	9. Sink or Float?	10. Watermelon and Grape	11. Is It Matter?	12. Snap Blocks	13. Back and Forth	14. When Is There Friction?	15. Marble Roll	16. Do the Waves Move the Boat?	17. Shadow Size	18. Rubber Band Box	19. Big and Small Magnets
chemical change					✓						
conservation of matter				✓							
describing motion							✓				
forces						✓					
friction						✓					
gases			✓								
light									✓		
liquids			✓								
magnetism											✓
magnets											✓
matter			✓								
motion							✓	✓			
parts and wholes				✓							
physical change					✓						
physical properties	✓	✓	✓								
pitch										✓	
shadows									✓		
sinking and floating	✓	✓									
solids			✓								
sound										✓	
states of matter			✓								
vibration										✓	
waves								✓			
weight				✓							

Sink or Float?

Circle the person with the best idea.

What are you thinking?

Sink or Float?

Teacher Notes

Purpose
The purpose of this assessment probe is to elicit children's ideas about physical properties. The probe is designed to find out if children recognize that an object or material that sinks can be made to float by changing its shape.

Related Concepts
sinking and floating, physical properties

Explanation
Bonita has the best answer: "I think sometimes clay sinks and sometimes it floats." Clay is denser than water. Therefore, by itself, clay sinks regardless of whether it is in the shape of a ball, hot dog, or pancake. However, clay can be flattened and the sides turned up into a boatlike form. This shape contains a volume of air (like an empty cup or boat), allowing the clay to float. It now contains two materials—clay and air. This means the mass-to-volume ratio (density) of the shape containing clay and air will be less than the mass-to-volume ratio of a clay shape that does contain air. While this explanation is for adult learners, the focus for K–2 should be on the change in shape, not the mass-to-volume ratio.

Curricular and Instructional Considerations for Grades K–2
Floating and sinking investigations provide a context for students to investigate and describe physical properties of objects such as size, shape, and weight. Students should have a variety of firsthand experiences in observing how objects of various sizes, shapes, and materials float or sink. While density is a concept taught in later grades, students at this level can begin to describe objects as being heavy or light for their size and recognize when something is shaped to be "hollow inside" or hold in air that helps it float. Making clay boats is a common activity at this level to show how changing the shape of an object can change its ability to float. Floating and sinking observations can be used in an engineering design context at this level to design and test models to determine how to use the properties of a material to design a floating object.

Administering the Probe
First, have students describe the properties of clay. If they are unfamiliar with clay, give them pieces of clay to feel. If enough clay is available, have students make different shapes out of clay before responding to the probe (ball, hot dog, and pancake as seen in the probe, but have them make other shapes as well), or have them describe different shapes they can make with clay. Have children circle or color the person with whom they agree the most and explain why they agree with that person. Explain to students that they should select the person whose idea best matches their thinking, not whose features they like most. See pages xxviii–xxxiii in the introduction for techniques used to guide "science talk" related to the probe.

Related Ideas in *Benchmarks for Science Literacy* (AAAS 2009)

K–2 Structure of Matter
• Objects can be described in terms of their properties. Some properties, such as hardness and flexibility, depend upon what

material the object is made of, and some properties, such as size and shape, do not.

Related Core Ideas in *A Framework for K–12 Science Education* (NRC 2012)

K–2 PS1.A: Structure and Properties of Matter

- Matter can be described and classified by its observable properties (e.g., visual, aural, textural), by its uses, and by whether it occurs naturally or is manufactured. Different properties are suited to different purposes.

Related *Next Generation Science Standards* (Achieve Inc. 2013)

K–2: Engineering Design

- K-2-ETS1-2: Develop a simple sketch, drawing, or physical model to illustrate how the shape of an object helps it function as needed to solve a given problem.

Grade 2: Matter and Its Interactions

- 2-PS1-1: Plan and conduct an investigation to describe and classify different kinds of materials by their observable properties.
- 2-PS1-2: Analyze data obtained from testing different materials to determine which materials have the properties that are best suited for an intended purpose.

Grade 5: Matter and Its Interactions

- 5-PS1-3: Make observations and measurements to identify materials based on their properties.

Related Research

- Students' ways of looking at floating and sinking include the roles played by mate-

rial, weight, shape, cavities, holes, air, and water (Driver, Squires, Rushworth, and Wood-Robinson 1994).
- A study conducted by Biddulph and Osborne (1984) asked students ages 7 to 14 why things float. The typical response was "because they are light."
- Some students use an intuitive rule of "More A–More B" (Stavy and Tirosh 2000). They reason that if you have a larger object (bigger piece of clay), then it must sink while the smaller piece will float.
- Children younger than age 5 typically ignore an object's size and focus on its "felt weight" (Smith, Carey, and Wiser 1984).

Suggestions for Instruction and Assessment

- Provide students with clay and have them make the shapes in the concept cartoon probe and drop them in water. Have them observe whether the shapes float or sink. Ask students if they can come up with a shape that will make the clay float, then test their ideas.
- After students have had an opportunity to investigate how they can make clay float, provide a real-world context. Give them a piece of aluminum foil, crush it into a ball, and drop it in water to observe how it sinks. Then have them shape another piece of aluminum foil into a flat-bottom boat and put it in water. Students will observe that it floats. Discuss why the boat floats and the ball sinks. Then give students some pieces of steel, such as nails, washers, or other pieces of hardware, and have them predict what happens when they are put in water. Have them test their ideas. Then show a picture of a steel ship such as an oil tanker floating on water, and ask students why the steel boat floats. How is this similar to the clay boat?

- Once students figure out how to change the shape of clay to make it float, turn this into an engineering activity by having students design, model, and test a clay boat that can hold the most marbles. Give each team of students the same amount of clay.

Related NSTA Resources

Keeley, P. 2010. Formative assessment probes: "More A–More B" rule. *Science and Children* 48 (2): 24–26.

Keeley, P., F. Eberle, and J. Tugel. 2007. *Uncovering student ideas in science, vol. 2: 25 more formative assessment probes.* Arlington, VA: NSTA Press.

Robertson, W. 2007. Science 101: How can an ocean liner made of steel float on water? *Science and Children* 44 (9): 54–59.

Smithenry, D., and J. Kim. 2010. Beyond predictions. *Science and Children* 48 (2): 48–52.

References

Achieve Inc. 2013. *Next generation science standards.* *www.nextgenscience.org/next-generation-science-standards.*

American Association for the Advancement of Science (AAAS). 2009. Benchmarks for science literacy online. *www.project2061.org/publications/bsl/online*

Biddulph, F., and R. Osborne. 1984. Pupil's ideas about floating and sinking. Paper presented to the Australian Science Education Research Association Conference, Melbourne, Australia.

Driver, R., A. Squires, P. Rushworth, and V. Wood-Robinson. 1994. *Making sense of secondary science: Research into children's ideas.* London: Routledge.

National Research Council (NRC). 2012. *A framework for K–12 science education: Practices, crosscutting concepts, and core ideas.* Washington, DC: National Academies Press.

Smith, C., S. Carey, and M. Wiser. 1984. A case study of the development of size, weight, and density. *Cognition* 21 (3): 177–237.

Stavy, R., and D. Tirosh. 2000. *How students (mis)understand science and mathematics: Intuitive rules.* New York: Teachers College Press.

Watermelon and Grape

The watermelon will sink.

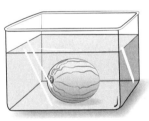

The grape will sink.

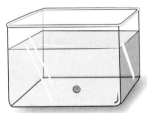

The watermelon will float.

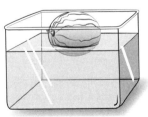

The grape will float.

What are you thinking?

Watermelon and Grape

Teacher Notes

Purpose

The purpose of this assessment probe is to elicit children's ideas about floating and sinking. The probe is designed to find out if children think size is the property that determines whether an object floats or sinks.

Related Concepts

sinking and floating, physical properties

Explanation

The best answers are "The watermelon will float" and "The grape will sink." Although the watermelon is much larger than the grape and its "felt weight" is much greater, its mass-to-volume ratio (density) is less than that of a grape. Also, its density is less than water; therefore it floats in water. The grape's mass-to-volume ratio is greater than the watermelon's even though its "felt weight" is much less. The density of a grape is greater than that of the water; therefore it sinks. Denser objects are "heavy for their size," while less-dense objects are "light for their size." An object denser than water sinks; an object less dense than water floats. It is the mass-to-volume ratio that makes a difference, not the size.

Curricular and Instructional Considerations for Grades K–2

At the K–2 level, floating and sinking provides a context for students to investigate and describe properties of objects such as size, shape, and weight. Students should have a variety of firsthand experiences in observing how objects of various sizes, shapes, weights, and materials float or sink. Although density is a concept that is taught in later grades, students at this level can begin to describe objects as being "heavy or light for their size." It is not important that students know whether a watermelon or grape floats or sinks, nor should K–2 students be expected to use the concept of density. What is important when using this probe is to determine whether students recognize that size alone does not determine whether an object floats or sinks and provide them with experiences to test this, such as big objects that float and small objects that sink, and vice versa.

Administering the Probe

Start off by asking the students if they have ever seen a watermelon and a grape. Make sure students are familiar with a watermelon and a grape before responding to the probe. If the actual objects are available, students can look at and feel them first. You can show pictures of the objects, preferably someone holding a watermelon so that students can get a sense of the size. You can also ask students to show you with their hands how big a watermelon and grape are before engaging with the probe. Ask students, "What do you think will happen when the watermelon and the grape are dropped in the water?" Make sure students know there are two parts to this probe. First, they select what they think will happen when the watermelon is put in the tub of water and put an *X* in or color the box. Then they make a prediction about the grape when it is placed in the water. *Note:* It is not important that students know the answer to this probe. What is important is the reasoning they use to support their prediction. See pages xxviii–xxxiii in the introduction for techniques used to guide "science talk" related to the probe.

Related Ideas in *Benchmarks for Science Literacy* (AAAS 2009)

K–2 Structure of Matter

- Objects can be described in terms of their properties. Some properties, such as hardness and flexibility, depend upon what material the object is made of, and some properties, such as size and shape, do not.

Related Core Ideas in *A Framework for K–12 Science Education* (NRC 2012)

K–2 PS1.A: Structure and Properties of Matter

- Matter can be described and classified by its observable properties (e.g., visual, aural, textural), by its uses, and by whether it occurs naturally or is manufactured.

Related *Next Generation Science Standards* (Achieve Inc. 2013)

Grade 2: Matter and Its Interactions

- 2-PS1-1: Plan and conduct an investigation to describe and classify different kinds of materials by their observable properties.

Grade 5: Matter and Its Interactions

- 5-PS1-3: Make observations and measurements to identify materials based on their properties.

Related Research

- Students' ways of thinking about floating and sinking include the roles played by material, weight, shape, cavities, holes, air,

and water (Driver, Squires, Rushworth, and Wood-Robinson 1994).

- A study conducted by Biddulph and Osborne (1984) asked students ages 7 to 14 why things float. The typical response was "because they are light."
- Some students use an intuitive rule of "More A-More B" (Stavy and Tirosh 2000). They reason that if you have a larger object, it must sink more.
- Children younger than age 5 typically ignore an object's size and focus on its "felt weight" (Smith, Carey, and Wiser 1984).
- Piaget's studies (1973) showed that children initially think of a pebble as being "light" but later describe it as "light for them" but "heavy for water." He showed that when children reach ages 9 and 10, they begin to relate the density of one material to that of another material by describing that some materials float because they are lighter than water (Driver, Squires, Rushworth, and Wood-Robinson 1994).

Suggestions for Instruction and Assessment

- A related probe, "Floating Logs," can be adapted for this grade level to find out if students think a larger object of the same material floats differently (Keeley, Eberle, and Tugel 2007).
- If a watermelon and grapes are available, this probe can be used as a P-E-O (Predict-Explain-Observe) probe to launch into an investigation (Keeley 2008). When students see that their observations do not match their predictions, they can make new claims and support them with the evidence from their investigation. Additionally, a variety of other large and small fruits or other objects can be used to have students make predictions about floating or sinking; explain the reason for their

predictions; and then test, observe, and revise their ideas.

- Extend children's investigations to objects of the same material but different sizes and shapes.

Related NSTA Resources

Keeley, P. 2010. Formative assessment probes: "More A-More B" rule. *Science and Children* 48 (2): 24–26.

Robertson, W. 2007. Science 101: How can an ocean liner made of steel float on water? *Science and Children* 44 (9): 54–59.

Smithenry, D., and J. Kim. 2010. Beyond predictions. *Science and Children* 48 (2): 48–52.

References

Achieve Inc. 2013. *Next generation science standards. www.nextgenscience.org/next-generation-science-standards.*

American Association for the Advancement of Science (AAAS). 2009. Benchmarks for science literacy online. *www.project2061.org/publications/bsl/online*

Biddulph, F., and R. Osborne. 1984. Pupil's ideas about floating and sinking. Paper presented to the Australian Science Education Research Association Conference, Melbourne, Australia.

Driver, R., A. Squires, P. Rushworth, and V. Wood-Robinson. 1994. *Making sense of secondary science: Research into children's ideas.* London: Routledge.

Keeley, P. 2008. *Science formative assessment: 75 practical strategies for linking assessment, instruction, and learning.* Thousand Oaks, CA: Corwin Press and Arlington, VA: NSTA Press.

Keeley, P., F. Eberle, and J. Tugel. 2007. *Uncovering student ideas in science, vol. 2: 25 more formative assessment probes.* Arlington, VA: NSTA Press.

National Research Council (NRC). 2012. *A framework for K–12 science education: Practices, crosscutting concepts, and core ideas.* Washington, DC: National Academies Press.

Piaget, J. 1973. *The child's conception of the world.* London: Paladin.

Smith, C., S. Carey, and M. Wiser. 1984. A case study of the development of size, weight, and density. *Cognition* 21 (3): 177–237.

Stavy, R., and D. Tirosh. 2000. *How students (mis)understand science and mathematics: Intuitive rules.* New York: Teachers College Press.

Is It Matter?

rock

ice cream

ice

house

soil

ant

water

air inside a balloon

sponge

dust

What are you thinking?

Is It Matter?

Teacher Notes

Purpose
The purpose of this assessment probe is to elicit children's ideas about the concept of matter. The probe is designed to find out what characteristics students use to describe the concept of matter.

Related Concepts
matter, states of matter, solids, liquids, gases

Explanation
Everything on the list is matter. Matter is made up of particles, has mass (and weight), and takes up space (has volume). It can exist as a solid, liquid, or gas (and, in some cases, as plasma). Matter is what makes up the materials, objects, or substances we encounter in our natural world and throughout the universe.

Curricular and Instructional Considerations for Grades K–2
Students begin to encounter the word *matter* in the primary grades, after they have had informal experiences exploring matter since the time they were infants. In school, they learn about different kinds of matter that exist, the states of matter, and the properties of matter. Therefore, it is important for students to develop and articulate a concept of matter before investigating matter and its interactions. Students develop a beginning notion of matter as the "stuff" that makes up objects and materials, beginning with matter they can see, feel, measure, and weigh (liquids and solids). Gases are more difficult, as students cannot see or "feel the weight" of gases. As students develop the idea that things are made of smaller parts and some parts are too small to see with our eyes, they begin to formulate a basic conceptual model of matter as being made of smaller particles. Although air is included on this list, it is not until grades 3–5 when students are expected to use a particle model to explain why air is matter.

Administering the Probe
Review the items on the list with students to make sure they are familiar with each object listed. Name each object as you associate it with the picture. This practice is especially important for English language learners. Make sure the pictures are clear to the students. If you have other pictures that illustrate each of the objects listed, you might show those to the students in addition to the ones on the student page. You can also provide some of the actual objects. For the balloon, make sure students know the probe is referring to the air inside the balloon, not the balloon itself. Instruct students to circle or color the things they think are matter. Additionally, you may ask them to put an *X* over the ones they do not think are matter. Have students explain the rule they used to decide whether the things on the list are considered matter, and listen carefully for the criteria they used to decide. See pages xxviii–xxxiii in the introduction for techniques used to guide "science talk" related to the probe.

Related Ideas in *Benchmarks for Science Literacy* (AAAS 2009)

K–2 Structure of Matter
- Objects can be described in terms of their properties. Some properties, such as hard-

ness and flexibility, depend upon what material the object is made of, and some properties, such as size and shape, do not.

Related Core Ideas in *A Framework for K–12 Science Education* (NRC 2012)

K–2 PS1.A: Structure and Properties of Matter

- Different kinds of matter exist (e.g., wood, metal, water), and many of them can be either solid or liquid.

3–5 PS1.A: Structure and Properties of Matter

- Matter of any type can be subdivided into particles that are too small to see, but even then the matter still exists and can be detected by other means.

Related *Next Generation Science Standards* (Achieve Inc. 2013)

Grade 2: Matter and Its Interactions

- 2-PS1-1: Plan and conduct an investigation to describe and classify different kinds of materials by their observable properties.

Grade 5: Matter and Its Interactions

- 5-PS1-3: Make observations and measurements to identify materials based on their properties.

Related Research

- Although in science classes the word *material* is used to designate any kind of matter or "stuff" that can be observed or detected in the world around us, children may initially use the word to mean those things that are required to make objects—for example, fabrics for clothes (Driver, Squires, Rushworth, and Wood-Robinson 1994).
- Although the word *stuff* may not be accepted as a scientific word, it has tangible connotations for students and therefore is useful for developing the idea that there are different kinds of "stuff" with different properties (Driver, Squires, Rushworth, and Wood-Robinson 1994).
- In a study conducted to find out what meaning students give to the word *matter*, 20% of middle-school-age students described matter as something tangible, meaning it can be handled and takes up space. By age 16, 66% of students described it this way (Bouma, Brandt, and Sutton 1990).
- Some students may accept solids as matter but not liquids (AAAS 2009).
- Having a correct conception of matter is necessary for students to understand ideas such as conservation of matter, mass, and weight (AAAS 2009).
- From an early age, children notice how objects differ in the way they appear to press down on the hands, shoulder, or head; they learn to "feel the weight" of objects. Children compare the "felt weight" of objects and generate the idea that "felt weight" is a characteristic property of the object, eventually using it to describe matter (Driver, Squires, Rushworth, and Wood-Robinson 1994).

Suggestions for Instruction and Assessment

- A related probe—"Is It Matter?"—can be adapted to use as an extension to this probe (Keeley, Eberle, and Farrin 2005).
- This probe can be used with the card-sort strategy (Keeley 2008) by having students work in small groups to sort each of the items placed on cards into three categories: things they think are matter, things they

do not think are matter, and things about which they are unsure. As students sort the cards, they must explain the reason why an object is or is not considered matter.

- Here are additional items you can use to challenge students' conception of matter: rotten apple (some students think a material is no longer matter when it is in a state of decay), cotton ball (some students think cotton is too soft to be matter), yogurt (some students will say it is neither a solid or a liquid so it cannot be matter), and various other materials and objects, such as sand, popcorn, a bird, melted butter, a crayon, a flower, and snow.

- Knowing students' conception of matter is a prerequisite to designing instruction regarding matter-related concepts.

- Do not assume students know what matter is when you begin using the terminology that is in the standards. Start by using the nonscientific, familiar operative word—*stuff*—until students are ready to understand and use the scientific word—*matter.*

- Be aware that using a definition such as "anything that has mass and occupies space" is meaningless to students if they do not know what mass is or what is meant by the term *occupying space* (volume). The term *mass* is intentionally not used in national standards at the elementary level. Although *weight* is not synonymous with *mass*, it is a stepping stone to the concept of mass that is used later, in middle school.

- Provide opportunities for students to weigh things that seem "weightless" to them, such as powders, dry sponges, and other "light" materials and objects.

- Have students experience what it means to take up space with solids and liquids. Although gases are more difficult at this age, students may see that air takes up space by blowing up a balloon or closing a zip-top bag that has air in it.

- Make sure students use the concept of matter in multiple contexts. If used only in a physical science context, students may not recognize matter as making up living things and Earth materials. Revisit the idea of matter in life science and Earth science contexts.

- Session 1 of the Annenberg video series *Essential Science for Teachers: Physical Science* has an excellent video clip of a second-grade teacher questioning her students about the concept of matter. The streaming video can be accessed at *www.learner.org/resources/series200.html#program_descriptions.*

Related NSTA Resources

Adams, B. 2006. Science shorts: All that matters. *Science and Children* 44 (1): 53–55.

Palmeri, A., A. Cole, S. DeLisle, S. Erickson, and J. Janes. 2008. What's the matter with teaching children about matter? *Science and Children* 46 (4): 20–23.

Royce, C. 2012. Teaching through trade books: What's the matter? *Science and Children* 49 (6): 22–24.

References

Achieve Inc. 2013. *Next generation science standards.* www.nextgenscience.org/next-generation-science-standards.

American Association for the Advancement of Science (AAAS). 2009. Benchmarks for science literacy online. *www.project2061.org/publications/bsl/online*

Bouma, H., I. Brandt, and C. Sutton. 1990. *Words as tools in science lessons.* Amsterdam: University of Amsterdam.

Driver, R., A. Squires, P. Rushworth, and V. Wood-Robinson. 1994. *Making sense of secondary science: Research into children's ideas.* London: Routledge.

Keeley, P. 2008. *Science formative assessment: 75 practical strategies for linking assessment, instruction,*

and learning. Thousand Oaks, CA: Corwin Press and Arlington, VA: NSTA Press.

Keeley, P., F. Eberle, and L. Farrin. 2005. *Uncovering student ideas in science, vol. 1: 25 formative assessment probes.* Arlington, VA: NSTA Press.

National Research Council (NRC). 2012. *A framework for K–12 science education: Practices, crosscutting concepts, and core ideas.* Washington, DC: National Academies Press.

Snap Blocks

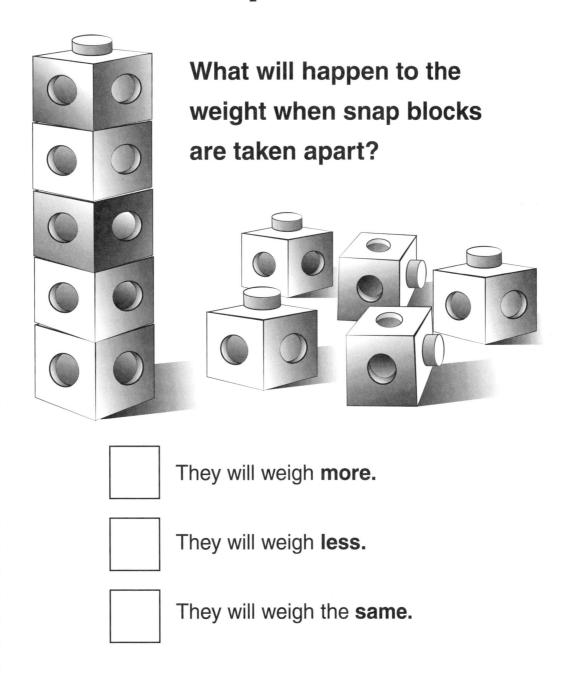

What will happen to the weight when snap blocks are taken apart?

☐ They will weigh **more.**

☐ They will weigh **less.**

☐ They will weigh the **same.**

What are you thinking?

Snap Blocks

Teacher Notes

Purpose
The purpose of this assessment probe is to elicit children's ideas about conservation of matter. The probe is designed to find out if students recognize that all the individual parts that make up an object taken together weigh the same as the whole object before it is taken apart.

Related Concepts
weight, conservation of matter, parts and wholes

Explanation
The best answer is "They will weigh the same." The total weight of the individual snap blocks taken together is equal to the weight of the snap blocks when they are connected to form one object. No new matter was added or taken away to change the weight. Conservation of matter is a scientific principle that applies to physical changes in objects, such as taking them apart, as well as chemical changes that result in the formation of new substances.

Curricular and Instructional Considerations for Grades K–2
Addressing conservation of matter with simple, everyday objects can begin in the primary grades. By the end of second grade, students develop the "parts and wholes" idea that things are made of parts. Although conserving matter is not a primary-level expectation and is not formally addressed until grades 3–5, students can begin to explore the property of weight by observing what happens to the weight when an object is taken apart. Primary students should have multiple experiences weighing objects and their parts before and after taking them apart. At this level, the term *mass* is not used.

Administering the Probe
Snap blocks are frequently used as manipulatives in mathematics. If you have snap blocks, model the probe scenario for students by showing them an object with snap blocks connected and then taken apart. If snap blocks are not available, substitute Lego blocks. Make sure students know they are comparing the weight of the snap blocks when they are all connected together to the total weight of all the individual snap blocks when they are taken apart and weighed together. Have students color or put an *X* in the box next to the claim that best matches their thinking about the weight of the snap blocks. See pages xxviii–xxxiii in the introduction for techniques used to guide "science talk" related to the probe.

Related Ideas in *Benchmarks for Science Literacy* (AAAS 2009)

3–5 Structure of Matter
- No matter how parts of an object are assembled, the weight of the whole object is always the same as the sum of the parts; and when an object is broken into parts, the parts have the same total weight as the original object.

Related Core Ideas in *A Framework for K–12 Science Education* (NRC 2012)

K–2 PS1.A: Structure and Properties of Matter

- A great variety of objects can be built up from a small set of pieces (e.g., blocks, construction sets). Objects or samples of a substance can be weighed, and their size can be described and measured.

3–5 PS1.A: Structure and Properties of Matter

- The amount (weight) of matter is conserved when it changes form, even in transitions in which it seems to vanish.

Related *Next Generation Science Standards* (Achieve Inc. 2013)

Grade 2: Matter and Its Interactions

- 2-PS1-3: Make observations to construct an evidence-based account of how an object made of a small set of pieces can be disassembled and made into a new object.

Grade 5: Matter and Its Interactions

- 5-PS1-2: Measure and graph quantities to provide evidence that regardless of the type of change that occurs when heating, cooling, or mixing substances, the total weight of matter is conserved.

Related Research

- Several studies have shown that the way a physical change is perceived may influence whether students regard material as being conserved during the change (Driver, Squires, Rushworth, and Wood-Robinson 1994).

- Stavy and Tirosh (2000) investigated intuitive rules used by students to explain conservation problems. Because the individual pieces look smaller than the whole objects, some children may reason that the weight of the pieces is less than the weight of the whole.

Suggestions for Instruction and Assessment

- A similar probe for elementary children that addresses conservation of matter using parts of an object is "Cookie Crumbles" (Keeley, Eberle, and Farrin 2005).

- This probe can be used as a P-E-O (Predict-Explain-Observe) probe (Keeley 2008). It can be followed up with an investigation in which students make a prediction and test their prediction by weighing the object made of snap blocks before taking it apart and weighing all the individual blocks together. Guide students to help them realize that all the parts together weigh the same as the whole. Students should also have an opportunity to test and compare the reverse—finding the total weight of all the individual parts and then putting them together and finding the weight of the whole object. However, note that some pan balances used with young children have pans large enough that if the objects are not placed the same distance from the fulcrum, they will not evenly balance, leading some children to think the weight changes.

- Have students count the number of snap blocks that are connected together to make the object. After taking the object apart, have students count the individual snap blocks. This should provide additional evidence by showing that there is the same number of blocks. No new blocks were added or taken away when the object was put together and taken apart. This helps students develop an

early precursor idea that the same amount of "stuff" or matter was there, which they can later (in grades 3–5) apply to substances and matter they cannot see.

- Provide students with opportunities to try out other "parts and wholes" by weighing a variety of objects and then weighing all of the objects' parts.

- Connect parts and wholes of objects to parts and wholes of numbers to make a link to the *Common Core State Standards, Mathematics.* A large number can be broken down into smaller numbers that equal the same as the larger number when added together (e.g., 6 = 2 + 4; 10 = 2 + 3 + 5).

References

Achieve Inc. 2013. *Next generation science standards.* www.nextgenscience.org/next-generation-science-standards.

American Association for the Advancement of Science (AAAS). 2001. *Atlas of science literacy.* (See "Conservation of Matter," pp. 56–57). New York: Oxford University Press.

American Association for the Advancement of Science (AAAS). 2009. Benchmarks for science literacy online. *www.project2061.org/publications/bsl/online*

Driver, R., A. Squires, P. Rushworth, and V. Wood-Robinson. 1994. *Making sense of secondary science: Research into children's ideas.* London: Routledge.

Keeley, P. 2008. *Science formative assessment: 75 practical strategies for linking assessment, instruction, and learning.* Thousand Oaks, CA: Corwin Press and Arlington, VA: NSTA Press.

Keeley, P., F. Eberle, and L. Farrin. 2005. *Uncovering student ideas in science, vol. 1: 25 formative assessment probes.* Arlington, VA: NSTA Press.

National Research Council (NRC). 2012. *A framework for K–12 science education: Practices, crosscutting concepts, and core ideas.* Washington, DC: National Academies Press.

Stavy, R., and D. Tirosh. 2000. *How students (mis)understand science and mathematics: Intuitive rules.* New York: Teachers College Press.

Back and Forth

Which changes can go back and forth?

☐ Ice to water.
Water to ice.

☐ Batter to cookie.
Cookie to Batter.

☐ Wood to ashes.
Ashes to wood.

☐ Hard butter to melted butter.
Melted butter to hard butter.

What are you thinking?

Back and Forth

Teacher Notes

Purpose

The purpose of this assessment probe is to elicit children's ideas about changes in matter. The probe is designed to reveal whether students recognize that not all changes are reversible.

Related Concepts

physical change, chemical change

Explanation

Ice to water, water to ice, hard butter to melted butter, and melted butter to hard butter are examples of changes that can go back and forth. These changes are reversible, meaning the matter can go back to its original form and material. These are physical changes in which the molecules that make up the substance do not change. Only the position and motion of the molecules relative to each other changed in these examples, as the state of matter changed from solid to liquid and back to solid again. When wood burns and forms ashes, a chemical change occurs. The molecules in the wood are rearranged to form new molecules in the ashes and gases given off during combustion. The ashes cannot be easily converted back into the wood because they are a different substance. When cookie dough is baked to form a cookie, a chemical change takes place. The cookie, a new material, cannot be converted back into cookie dough, a different material. Physical changes are generally easier than chemical changes to reverse (return to original conditions).

Curricular and Instructional Considerations for Grades K–2

In the primary grades, students observe and describe a variety of changes in matter. They should subject materials to treatments such as mixing, heating, freezing, cutting, wetting, dissolving, bending, and exposing to light to see how the materials change and observe whether any of the changes are reversible. Students should have opportunities to observe how some changes result in new materials that are different from the original materials and that they cannot reverse the change and go back to the original material (e.g., burning paper, baking a cake). This is a precursor idea that will help students distinguish between physical and chemical changes in later grades.

Administering the Probe

Go over each of the pictures to make sure students know what the pictures represent. Make sure students are familiar with each of the materials and understand that the question is aimed at whether they think they can get the same material back again after the change. For example, when ice is melted to form water, can the water be turned back into ice? If possible, show students an example or examples of an original and changed material. Have students put an X in or color the box or boxes they think show a change in matter that can go back and forth between different forms. See pages xxviii–xxxiii in the introduction for techniques used to guide "science talk" related to the probe.

Related Ideas in *Benchmarks for Science Literacy* (AAAS 2009)

K–2 Structure of Matter

- Things can be done to materials to change some of their properties, but not all materials respond the same way to what is done to them.

3–5 Structure of Matter

- Heating and cooling can cause changes in the properties of materials, but not all materials respond the same way to being heated and cooled.

Related Core Ideas in *A Framework for K–12 Science Education* (NRC 2012)

K–2 PS1.B: Chemical Reactions

- Heating or cooling a substance may cause changes that can be observed. Sometimes these changes are reversible (e.g., melting and freezing), and sometimes they are not (e.g., baking a cake, burning fuel).

3-5 PS1.B: Chemical Reactions

- When two or more substances are mixed, a new substance with different properties may be formed; such occurrences depend on the substances and the temperature.

Related *Next Generation Science Standards* (Achieve Inc. 2013)

Grade 2: Matter and Its Interactions

- 2-PS1-4: Construct an argument with evidence that some changes caused by heating or cooling can be reversed and some cannot.

Grade 5: Matter and Its Interactions

- 5-PS1-3: Make observations and measurements to identify materials based on their properties.
- 5-PS1-4: Conduct an investigation to determine whether the mixing of two or more substances results in new substances.

Related Research

- Stavridou and Solomonidou (1989) explored Greek students' (ages 8 to 17) conceptions of physical and chemical changes. The researchers' general conclusion was that children who used the reversibility criterion (ability to go back and forth between different forms) were able to distinguish between chemical and physical changes, whereas those who used other criteria failed to do so. Students who used the reversibility criterion considered a reversible change to be nonradical and therefore physical. Although many chemical reactions are reversible, children who used the reversibility criterion considered chemical changes to be irreversible (Driver, Squires, Rushworth, and Wood-Robinson 1994).
- Several studies have found that children often use the term *chemical change* to encompass changes in physical state and other physical transformations (Driver, Squires, Rushworth, and Wood-Robinson 1994).
- How well students make a distinction between physical and chemical changes depends partly on their conception of what a substance is. For example, if they regard ice as a different substance than water, then they are likely to regard the melting of ice as a chemical change (Driver, Squires, Rushworth, and Wood-Robinson 1994).

Suggestions for Instruction and Assessment

- Provide multiple opportunities for children to experience reversible changes such as changing from solid to liquid and back to solid, taking things apart and putting them back together, changing the shape of a material such as clay, and folding and unfolding.
- Provide opportunities for children to observe changes they cannot reverse, such as cooking a raw egg, making toast, and rusting a nail.
- Connect the idea of reversible changes during a change of state with the idea that liquid water and ice are the same substance.

Related NSTA Resources

Adams, B. 2006. Science shorts: All that matters. *Science and Children* 44 (1): 53–55.

Ashbrook, P. 2006. The early years: The matter of melting. *Science and Children* 43 (4): 18–21.

Palmeri, A., A. Cole, S. DeLisle, S. Erickson, and J. Janes. 2008. What's the matter with teaching children about matter? *Science and Children* 46 (4): 20–23.

References

Achieve Inc. 2013. *Next generation science standards.* *www.nextgenscience.org/next-generation-science-standards.*

American Association for the Advancement of Science (AAAS). 2009. Benchmarks for science literacy online. *www.project2061.org/publications/bsl/online*

Driver, R., A. Squires, P. Rushworth, and V. Wood-Robinson. 1994. *Making sense of secondary science: Research into children's ideas.* London: Routledge.

National Research Council (NRC). 2012. *A framework for K–12 science education: Practices, crosscutting concepts, and core ideas.* Washington, DC: National Academies Press.

Stavridou, H., and C. Solomonidou. 1989. Physical phenomena-chemical phenomena: Do students make the distinction? *International Journal of Science Education* 11 (1): 83–92.

When Is There Friction?

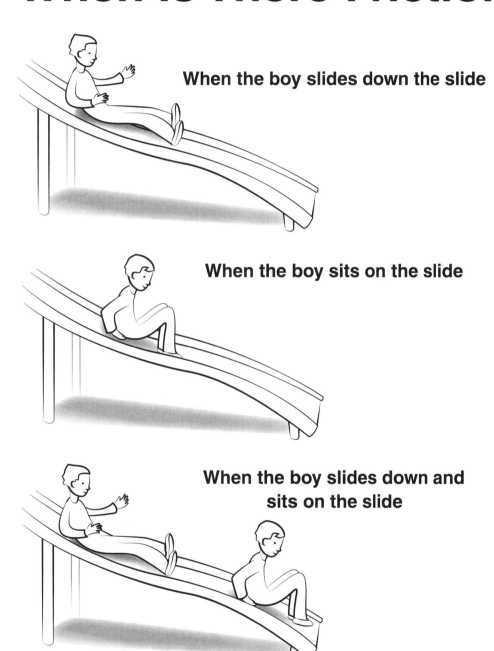

When the boy slides down the slide

When the boy sits on the slide

When the boy slides down and sits on the slide

What are you thinking?

When Is There Friction?

Teacher Notes

Purpose

The purpose of this assessment probe is to elicit children's ideas about friction. The probe is designed to find out whether students recognize that friction can act on a nonmoving object as well as on a sliding object.

Related Concepts

friction, forces

Explanation

The best answer is "When the boy slides down and sits on the slide." Interactions between objects can be explained with forces. *A Framework for K–12 Science Education* states, "An object sliding on a surface or sitting on a slope experiences a pull due to friction on the object due to the surface that opposes the object's motion" (NRC 2012, p. 115). The boy sliding down the slide is an example of kinetic friction, in which the surfaces (the boy's pants and the surface of the slide) are sliding past each other. The boy sitting on the slide is an example of static friction where the surfaces are not moving relative to each other. Static friction is what allows the boy to sit on the slide without sliding down it—that is, it is the force that resists the movement of one object against another when the objects are initially at rest. Another example would be a box that slides down a smooth, wood ramp versus a box that sits on a carpet-covered ramp.

Curricular and Instructional Considerations for Grades K–2

Prior to the release of *A Framework for K–12 Science Education* (NRC 2012), the concept of friction was not addressed until fifth grade or later. Introducing it earlier in the elementary curriculum is a curricular change for many teachers. At the K–2 level, it is not important that students know the names of the different types of friction. Rather, the emphasis should be on a variety of observational experiences that allow students to see how the interaction between a surface and an object affects the object's motion. Observations can be used in engineering design activities. *Note:* The concept of static friction is explicitly included as a K–2 disciplinary core idea in *A Framework for K–12 Science Education*; however, it is not included as an explicit disciplinary core idea in the *NGSS*.

Administering the Probe

This probe is best used after students have encountered the basic idea of friction and are familiar with the word. Ask students if they have ever slid down a slide. What was it like when they went fast down the slide? Was there ever a time when they had difficulty sliding down the slide or got stuck on the slide? Consider modeling the probe scenario for students with a block and a ramp that the block can sit on (carpeted, foam, or sandpaper surface) and a ramp that the block slides down (smooth, wood surface). Instruct students to color or circle the picture that best matches their ideas about friction. Check to make sure they know what each drawing represents. See pages xxviii–xxxiii in the introduction for techniques used to guide "science talk" related to the probe.

Related Ideas in *Benchmarks for Science Literacy* (AAAS 2009)

K–2 Motion

- The way to change how something is moving is to give it a push or pull.

Related Core Ideas in *A Framework for K–12 Science Education* (NRC 2012)

K–2 PS2.A: Forces and Motion

- An object sliding on a surface or sitting on a slope experiences a pull due to friction on the object due to the surface that opposes the object's motion.

3–5 PS2.A: Forces and Motion

- Each force acts on one particular object and has both a strength and direction. An object at rest typically has multiple forces acting on it, but they add to give zero net force on the object.

Related *Next Generation Science Standards* (Achieve Inc. 2013)

Kindergarten: Motion and Stability: Forces and Interactions

- K-PS2-1: Plan and conduct an investigation to compare the effects of different strengths or different directions of pushes and pulls on the motion of an object.
- K-PS2-2: Analyze data to determine if a design solution works as intended to change the speed or direction of an object with a push or a pull.

Grade 3: Motion and Stability: Forces and Interactions

- 3-PS2-1: Plan and conduct an investigation to provide evidence of the effects of balanced and unbalanced forces on the motion of an object.

Related Research

- In a study by Stead and Osborne (1980), some students thought that if a box is motionless on a slope, there is no friction because there is no detectable rubbing, heat, or wearing down of surfaces.
- Stead and Osborne (1981) reported that in their study of children's ideas about friction, 50% of the 13-year-olds associated friction with rubbing.
- Other ideas about friction are that it depends on movement; is associated with energy, especially heat; occurs only between solids; and "does this and that" as though it were an object (Driver, Squires, Rushworth, and Wood-Robinson 1994).

Suggestions for Instruction and Assessment

- A similar probe, "Friction," can be used with modification after students have had an opportunity to develop the core ideas related to friction (Keeley and Harrington 2010).
- Students can use their push and pull observations of objects sliding down or resting on a ramp in an engineering design context. For example, students can use models to design "sliding mats" to determine which type of material would be best to sit on when sliding down a slide. Then they can be challenged to design a surface for a steep ramp on which a person could sit without sliding down.
- Have students investigate sliding the same object down three different ramps. Contrast an object that slides easily down a

ramp surface with a ramp surface that does not let the object slide easily, and then a surface where the object does not slide at all, to illustrate the difference in the friction between different types of surfaces and the object. Additionally, have students investigate objects of similar shape but made of different materials as they are placed on a steep ramp. Have students observe which objects slide down the ramp and which ones "sit" on the ramp without moving, tying their observations to the idea of friction between the object and the type of surface with which the object is in contact.

- Encourage students to come up with other examples of friction involving moving and stationary objects. For example, what could students sit on, wear, or do to their clothes to sit on a playground slide without sliding down the slide or sliding really fast?
- Contrast "sticking" a piece of felt to a felt board with trying to stick the felt to the wall or a whiteboard. What other kinds of things "stick" or "slide off"? Contrast giving an object a push on a bare floor so it slides as opposed to a push on a carpeted floor.

Related NSTA Resources

Konicek-Moran, R. 2011. *Yet more everyday science mysteries: Stories for inquiry-based science teaching.* (See "Stuck," pp. 183–191.) Arlington, VA: NSTA Press.

Robertson, W. 2002. *Force and motion: Stop faking it! Finally understanding science so you can teach it.* Arlington, VA: NSTA Press.

References

Achieve Inc. 2013. *Next generation science standards.* www.nextgenscience.org/next-generation-science-standards.

American Association for the Advancement of Science (AAAS). 2009. Benchmarks for science literacy online. *www.project2061.org/publications/bsl/online*

Driver, R., A. Squires, P. Rushworth, and V. Wood-Robinson. 1994. *Making sense of secondary science: Research into children's ideas.* London: Routledge.

Keeley, P., and R. Harrington. 2010. *Uncovering student ideas in physical science, vol. 1: 45 new force and motion assessment probes.* Arlington, VA: NSTA Press.

National Research Council (NRC). 2012. *A framework for K–12 science education: Practices, crosscutting concepts, and core ideas.* Washington, DC: National Academies Press.

Stead, K., and R. Osborne. 1980. *Friction.* LISP working paper 19. Hamilton, New Zealand: University of Waikato, Science Education Research Institute.

Stead, K., and R. Osborne. 1981. What is friction? Some children's ideas. *New Zealand Science Teacher* 27: 51–57.

Marble Roll

How will the marble roll on to the floor?

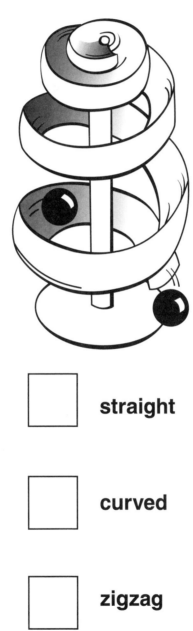

☐ straight

☐ curved

☐ zigzag

What are you thinking?

Marble Roll

Teacher Notes

Purpose

The purpose of this assessment probe is to elicit children's descriptions of motion. The probe is designed to reveal how students describe the path of a moving object as it leaves a winding track.

Related Concepts

motion, describing motion

Explanation

The best answer is "straight," and students should draw a straight line. The marble will travel in a straight line when it leaves the track. As the marble rolls down the marble tower's spiral track, a force toward the center of the spiral (a centripetal force) caused by the outside wall of the track keeps the marble rolling in a spiral path. When the marble leaves the end of the track, it is no longer in contact with the walls of the track. Without the track pushing on it, the marble no longer has a center-directed force acting on it that causes it to roll in a curved path. According to Newton's first law of motion, an object will remain at rest or in uniform motion in a straight line unless acted on by an outside force. There is no longer a center-directed force exerted by the wall of the track pushing on the marble, so the marble rolls off the track and across the floor in a straight path. It will continue this way unless an outside force causes it to change direction or slow down and stop.

Curricular and Instructional Considerations for Grades K–2

Students at the primary level may have played with curved marble towers, winding chutes, or toy cars moving on curved tracks as well as a variety of other moving objects that travel in curved paths. Their experiences during play, as well as in school, help them develop descriptions of motion that include straight, zigzag, curved, round and round, back and forth, up and down, and fast and slow. At this level, students are not expected to know and use Newton's first law. What is important is that they make predictions and observations of motion and learn words that can be used to describe different motions.

Administering the Probe

Make sure students understand the context of the probe. If you have a marble track or other type of curved track or flexible tubing, show it to students so they can see the path the marble takes on the tower. Make sure students know they should describe how they think the marble will move across the floor when it leaves the end of the track. Have students color or put an *X* in the box that best matches their thinking, then draw a picture of the path next to the word they chose. You might consider reviewing motion words first and drawing paths that show what the motion looks like before students make a prediction using a motion word. See pages xxviii–xxxiii in the introduction for techniques used to guide "science talk" related to the probe.

Related Ideas in *Benchmarks for Science Literacy* (AAAS 2009)

K–2 Motion

- Things move in many different ways, such as straight, zigzag, round and round, back and forth, and fast and slow.
- The way to change how something is moving is to give it a push or a pull.

3–5 Motion

- Changes in speed or direction of motion are caused by forces.

Related Core Ideas in *A Framework for K–12 Science Education* (NRC 2012)

K–2 PS2.A: Forces and Motion

- Pushing or pulling on an object can change the speed or direction of its motion and can start or stop it.

3–5 PS2.A: Forces and Motion

- The patterns of an object's motion in various situations can be observed and measured; when past motion exhibits a regular pattern, future motion can be predicted from it.

Related *Next Generation Science Standards* (Achieve Inc. 2013)

Kindergarten: Motion and Stability: Forces and Interactions

- K-PS2-1: Plan and conduct an investigation to compare the effects of different strengths or different directions of pushes and pulls on the motion of an object.

Grade 3: Motion and Stability: Forces and Interactions

- 3-PS2-2: Make observations and/or measurements of an object's motion to provide evidence that a pattern can be used to predict future motion.

Related Research

- Students often expect that objects moving in a curved path because of a wall or constraint will continue to do so when the wall or constraint is removed. This belief that the wall or constraint "trains" the object to follow a curved path is deeply rooted in students and persists even with targeted instruction in middle and high school (Arons 1997).
- Students have difficulty perceiving the direction of motion in a straight line when they encounter situations such as an object set in motion inside a curved hollow tube (Gunstone and Watts 1985).
- Children need to develop the language tools to describe motion appropriately prior to developing an understanding of the principles (Driver, Squires, Rushworth, and Wood-Robinson 1994).

Suggestions for Instruction and Assessment

- A similar probe, "Rolling Marbles," can be adapted for this grade level (Keeley, Eberle, and Dorsey 2008).
- Provide students with a variety of motions to observe to help them develop the vocabulary needed to describe motion. Build a word wall of motion descriptions.
- This probe can be used as a P-E-O (Predict-Explain-Observe) probe for the purpose of having students predict, explain, and observe the motion and then revise their description based on the observation (Keeley 2008). If a marble tower is not available, use a thick piece of clear, flexible

tubing. Have students help hold the tubing so it has a winding shape. Place a small marble or other round object into the tubing and have students observe it as it travels through the winding tube. Students are typically surprised to find the marble travels in a straight line when it exits the tube or marble tower.

- Encourage students to share their experiences with toy race car tracks or other toys that involve circular motions and describe what happens when an object leaves the winding path.
- In addition to solids, students can be challenged to predict and describe the motion and path of liquids. If an outdoor hose and spigot is available, challenge students to predict what the path of water will be when it comes out of a hose that has been coiled up. Encourage them to describe how the path is similar to the marble on the curved track.

Related NSTA Resources

American Association for the Advancement of Science (AAAS). 2001. *Atlas of science literacy.* Vol. 1. (See "Laws of Motion Map," pp. 62–63.) New York: Oxford University Press.

Ashbrook, P. 2008. The early years: Objects in motion. *Science and Children* 45 (7): 14–16.

Harris, J. 2004. Science 101: Are there different types of force and motion? *Science and Children* 41 (6): 19.

Robertson, W. 2002. *Force and motion, Stop faking it! Finally understanding science so you can teach it.* Arlington, VA: NSTA Press.

Stein, M. 1998. Toying with science. *Science and Children* 36 (1): 35–39.

References

Achieve Inc. 2013. *Next generation science standards. www.nextgenscience.org/next-generation-science-standards.*

American Association for the Advancement of Science (AAAS). 2009. Benchmarks for science literacy online. *www.project2061.org/publications/bsl/online*

Arons, A. 1997. *Teaching introductory physics.* New York: John Wiley and Sons.

Driver, R., A. Squires, P. Rushworth, and V. Wood-Robinson. 1994. *Making sense of secondary science: Research into children's ideas.* London: Routledge.

Gunstone, R., and M. Watts. 1985. Force and motion. In *Children's ideas in science,* ed. R. Driver, E. Guesne, and A. Tiberghien, pp. 85–104. Milton Keynes, UK: Open University Press.

Keeley, P. 2008. *Science formative assessment: 75 practical strategies for linking assessment, instruction, and learning.* Thousand Oaks, CA: Corwin Press and Arlington, VA: NSTA Press.

Keeley, P., F. Eberle, and C. Dorsey. 2008. *Uncovering student ideas in science, vol. 3: Another 25 formative assessment probes.* Arlington, VA: NSTA Press.

National Research Council (NRC). 2012. *A framework for K–12 science education: Practices, crosscutting concepts, and core ideas.* Washington, DC: National Academies Press.

Do the Waves Move the Boat?

☐ **Yes, the boat moves with the waves.**

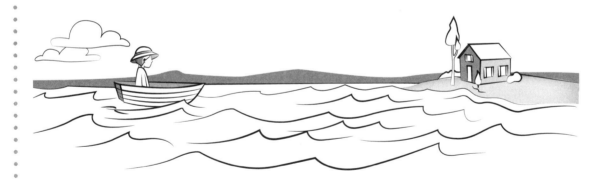

☐ **No, the boat does not move with the waves.**

What are you thinking?

Do the Waves Move the Boat?

Teacher Notes

Purpose

The purpose of this assessment probe is to elicit children's ideas about waves. The probe is designed to reveal students' ideas about water waves and the waves' interactions with floating objects.

Related Concepts

waves, motion

Explanation

The best answer is "No, the boat does not move with the waves." A wave is a pattern of motion resulting from a disturbance that travels through a medium from one location to another. In this case, the medium is the water. The wave moves through the water (medium), but the water is not moving with the wave. In other words, the water (with the boat on it) is not carried toward the shore by the waves. A wave transports energy and does so without transporting matter in the direction of the wave (e.g., the water in the lake). If you were to observe the boat on the lake, you would notice that it gently bobs up and down as the waves pass by it. The boat stays in about the same position and is not carried toward the shore because the water it is floating on is not transported by the waves. An analogy would be people in a football stadium doing the wave. The wave travels around the stadium, but the people stay at their seats and do not move with the wave.

Curricular and Instructional Considerations for Grades K–2

Prior to the release of *A Framework for K–12 Science Education* (NRC 2012), the concept of waves was not addressed until middle school. Introducing it earlier in the elementary curriculum is a curricular change for many teachers. At the K–2 level, it is not important that students know about different types of waves, wave energy, and wave terminology. Rather, the emphasis should be on a variety of observational experiences with water waves that allow students to see the patterns of motion in the water. Further observations and testing can reveal that the water is not moving with the waves. The energy explanation should wait until later grades, when students learn about the concept of energy and how energy is transferred. (*Note:* Since the release of the *NGSS*, the *Framework*'s K–2 disciplinary core idea related to water waves not carrying an object has been moved to the grades 3–5 band in the *NGSS*.)

Administering the Probe

This is one of the more challenging probes in this collection. Check first to see if students have seen or are familiar with waves on water. It is important to describe the scenario of the probe so that students are clear about the context. The boat is out in a lake, quite a distance from the shore. Point to the boat and its location in both pictures. There are no

currents in the lake, but there are waves moving on the lake. Point to the waves in both pictures. Explain that these waves are out in the lake and not breaking near the shore. (Some students confuse water movement and waves from their experiences on the shoreline when waves break and the water moves forward.) Ask students if they think the waves will carry the boat toward the shore—that is, do they think the boat rides on the waves as the waves move? Do the waves move the boat toward the shore? Or do they think the boat will stay in about the same place and not be carried by the waves? However you decide to describe it, be sure children distinguish between the waves carrying things that are on the water versus the waves passing by the things that are on the water. It is a difficult concept to describe to young children, especially if they have never experienced seeing waves on water. Point out the difference in location in the two pictures. Explain that the first picture shows the boat on the lake moved closer to the shore. Explain that the second picture shows the boat has not moved further toward the shore. Explain to the students that if they choose yes, that means they think the waves carry the boat and move it along. If they choose no, that means they think the waves do not carry the boat and move it along, but the boat stays in about the same place and the waves move past the boat. See pages xxviii–xxxiii in the introduction for techniques used to guide "science talk" related to the probe.

Related Ideas in *Benchmarks for Science Literacy* (AAAS 2009)

Waves are not addressed in the *Benchmarks* until middle school.

Related Core Ideas in *A Framework for K–12 Science Education* (NRC 2012)

K–2 PS4.A: Wave Properties

- Waves, which are regular patterns of motion, can be made in water by disturbing the surface. When waves move across the surface of deep water, the water goes up and down in place; it does not move in the direction of the wave—observe, for example, a bobbing cork or seabird—except when the water meets the beach.

3–5 PS4.A: Wave Properties

- Waves of the same type can differ in amplitude (height of the wave) and wavelength (spacing between wave peaks). Waves can add to or cancel out one another as they cross, depending on their relative phase.

Related *Next Generation Science Standards* (Achieve Inc. 2013)

Grade 4: Waves and Their Applications in Technologies for Information Transfer

- 4-PS4-1: Develop a model of waves to describe patterns in terms of amplitude and wavelength and that waves can cause objects to move.

Related Research

- As of this book's publication, there are no research articles describing children's conceptions of water wave phenomena. This is an area that needs further research.
- Informal interviews with young children during the field test of this probe indicated they thought waves moved objects and that a boat moved on a wave. Be aware that

students also confused this idea with surfers and riding waves at the beach.

Suggestions for Instruction and Assessment

- Provide students with opportunities to observe how waves form outward from a disturbance, such as tossing pebbles into a pool of water and observing the ripples (small waves) that form.

- Provide students with opportunities to see that the water does not move with the waves. This can be done by floating a small piece of Styrofoam in a kiddie pool or tub of water. Drop an object into the water to create waves that move outward. Have students observe how the Styrofoam does not get carried by the waves to the edge of the pool or tub but bobs on the waves as they pass by.

Related NSTA Resources

Adams, B. 2007. Science shorts: Making waves. *Science and Children* 44 (5): 50–52.

American Association for the Advancement of Science (AAAS). 2001. *Atlas of science literacy.* Vol. 1. (See "Waves" map.) Washington, DC: AAAS.

References

Achieve Inc. 2013. *Next generation science standards.* www.nextgenscience.org/next-generation-science-standards.

American Association for the Advancement of Science (AAAS). 2009. Benchmarks for science literacy online. *www.project2061.org/publications/bsl/online*

National Research Council (NRC). 2012. *A framework for K–12 science education: Practices, crosscutting concepts, and core ideas.* Washington, DC: National Academies Press.

Shadow Size

How can you make the shadow bigger?

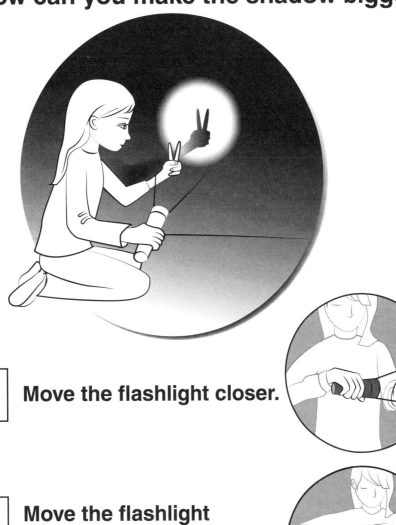

☐ Move the flashlight closer.

☐ Move the flashlight
farther away.

☐ It doesn't matter. The shadow stays the same.

What are you thinking?

Shadow Size

Teacher Notes

Purpose

The purpose of this assessment probe is to elicit children's ideas about how light interacts with an object to form a shadow. The probe is designed to reveal children's ideas about the size of a shadow in relation to the distance from a light source.

Related Concepts

light, shadows

Explanation

The best answer is "Move the flashlight closer." The smaller the distance between a light source and an object, the larger the shadow; the larger the distance between a light source and an object, the smaller the shadow.

Curricular and Instructional Considerations for Grades K–2

Observing changes in shadow size, direction, and shape is a common curricular activity that helps students develop ideas about how light interacts with objects. At this level, the investigations are observational. Students should be encouraged to analyze data and look for patterns that can be used to explain relationships between the position of a light source and an object.

Administering the Probe

Make sure students can explain what a shadow is before using the probe. Demonstrate how to form a shadow with your hand and a flashlight so that students understand the probe scenario (but do not show changes in shadow size by moving the distance between your hand and the light source). Make sure students know that the distance from the hand to the shadow on the wall does not change. It is the distance from the flashlight to the hand when the lights are turned off that students are asked to consider. Without turning on the flashlight, model what it means to move the flashlight closer to the hand or farther from the hand, or have it stay the same. Read each answer choice with the children to make sure they understand it and ask them to color or put an *X* in the box that best matches their ideas. See pages xxviii–xxxiii in the introduction for techniques used to guide "science talk" related to the probe.

Related Ideas in *Benchmarks for Science Literacy* (AAAS 2009)

3–5 Motion

- Light travels and tends to maintain its direction of motion until it interacts with an object or material. Light can be absorbed, redirected, bounced back, or allowed to pass through.

Related Core Ideas in *A Framework for K–12 Science Education* (NRC 2012)

K–2 PS4.B: Electromagnetic Radiation

- Some materials allow light to pass through them, others allow only some light through, and others block all light and create a dark shadow on any surface behind

them (i.e., on the other side from the light source), where the light cannot reach.

Related *Next Generation Science Standards* (Achieve Inc. 2013)

Grade 1: Waves and Their Applications in Technologies for Information Transfer

- 1-PS4-3: Plan and conduct an investigation to determine the effect of placing objects made with different materials in the path of a beam of light.

Grade 4: Waves and Their Applications in Technologies for Information Transfer

- 4-PS4-2: Develop a model to describe that light reflecting from objects and entering the eye allows objects to be seen.

Related Research

- Students' preconceptions about shadows come from their everyday personal experiences, even though these experiences may vary from student to student. These preconceptions, based on students' experiences, can put learning constraints on students' development of an understanding of light and shadows (Barrows 2012)
- Studies by Neal, Smith, and Johnson (1990) showed that some students do not connect shadows with a light source. Other students consider a shadow as being pushed out by the light. The researchers also found that some students think the size of a shadow is based on the size of the object.
- Some students view shadows as objects rather than understanding that shadows are created when light is blocked. Conceptual development is required for students to understand the relationship between a light source, an object, and the shadow cast by the object. Working with flashlights can

provide children an opportunity to directly challenge their everyday conceptions about shadows, providing them with powerful early experiences in scientific ways of knowing (Magnusson and Palincsar 2005).

Suggestions for Instruction and Assessment

- Another probe on shadows, "Me and My Shadow," can be adapted for this grade level (Keeley, Eberle, and Dorsey 2008). This probe examines the relationship between the position of the Sun and the length of a shadow.
- This probe can be used as a P-E-O (Predict-Explain-Observe) probe by having students predict how they should move the flashlight to make an object's shadow bigger, explain why they think the shadow will get bigger, then test their idea and make observations (Keeley 2008). If students' observations do not match their predictions, they can revise the probe and state a new claim, supported with evidence from the investigation.
- Instead of moving the flashlight, position a flashlight in a fixed position and have students predict how they can move an object to make its shadow smaller. Cut out a simple shape (e.g., heart, star, or triangle) and tape it on a popsicle stick to use for the object. Use a marked piece of tape to show equal intervals in the distance between the flashlight and the object. Place white paper against the wall. Starting with the first interval closest to the flashlight, have students trace the cast shadow on the paper. Then repeat with a new sheet of paper for the next interval, and so on. Have students display the data (their tracings) showing the relationship between distance from the light source and the size of the shadow. Have them explain what the data show.
- After students have discovered the relationship between the size of an object's shadow

and the object's distance from a light source, have them observe and describe what happens to the shape and size of a shadow when the flashlight shines on the object at different angles.

- After students have discovered how to change the size of a shadow, have them make shadow puppets by creating figures of people, animals, and objects attached to popsicle or paint sticks and stage a puppet show in which the sizes of the shadows change throughout the story. Have students explore how different cultures use shadow puppets. For instance, the oldest forms of puppetry were shadow puppets used in India and China more than 1,000 years ago.

Related NSTA Resources

Barrows, L. 2007. Bringing light onto shadows. *Science and Children* 44 (9): 43–45.

Carrejo, D., and J. Reinhartz. 2012. Science shorts: Modeling light and shadows. *Science and Children* 50 (2): 80–83.

References

Achieve Inc. 2013. *Next generation science standards. www.nextgenscience.org/next-generation-science-standards.*

American Association for the Advancement of Science (AAAS). 2009. Benchmarks for science literacy online. *www.project2061.org/publications/bsl/online*

Barrows, L. 2012. Helping students construct understanding about shadows. *Journal of Education and Learning* 1 (2): 188–191.

Keeley, P. 2008. *Science formative assessment: 75 practical strategies for linking assessment, instruction, and learning.* Thousand Oaks, CA: Corwin Press and Arlington, VA: NSTA Press.

Keeley, P., F. Eberle, and C. Dorsey. 2008. *Uncovering student ideas in science, vol. 3: Another 25 formative assessment probes.* Arlington, VA: NSTA Press.

Magnusson, S., and A. Palincsar. 2005. Teaching to promote the development of scientific knowledge and reasoning about light at the elementary school level. In *How students learn science in the classroom,* ed. S. Donovan and J. Bransford, pp. 421–469. Washington, DC: National Academies Press.

National Research Council (NRC). 2012. *A framework for K–12 science education: Practices, crosscutting concepts, and core ideas.* Washington, DC: National Academies Press.

Neal, D., D. Smith, and V. Johnson. 1990. Implementing conceptual change teaching in primary science. *The Elementary School Journal* 91: 109–131.

Rubber Band Box

What are you thinking?

Rubber Band Box

Teacher Notes

Purpose

The purpose of this assessment probe is to elicit children's ideas about variations in sound. The probe is designed to reveal students' thinking about factors that affect pitch.

Related Concepts

sound, vibration, pitch

Explanation

Wade has the best answer: "I think the thick rubber bands make the lowest sound." Sound is produced by a vibrating object. In this example, the vibrating object is the rubber band that is stretched across the box and plucked. What our ears perceive as the "highness" or "lowness" of the sound produced by the vibrating rubber bands is called the pitch. The more vibrations per second (frequency), the higher the pitch. Because the thick rubber band vibrates at a lower frequency than the thin rubber band, its pitch is lower.

Curricular and Instructional Considerations for Grades K–2

At the primary level, students learn how sound is produced and have opportunities to examine and describe different types of sounds and their sources. They should have opportunities to observe the vibrations of different types of sound-producing objects and compare the loudness and pitch of different sounds. They should also have opportunities to observe how sound can make objects vibrate. Students at this level often learn about sound in the context of musical instruments. As a result, students may be limited to thinking only objects similar to musical instruments produce sound through vibration. It is important to ensure that curriculum materials and activities develop the generalization that all sounds are produced by vibrating matter. Students can also engage in engineering design activities in which they apply their knowledge about sound and vibration.

Administering the Probe

First, model a low sound and a high sound for students so they have an operational understanding of pitch. Then show students how you can make a box guitar by stretching three rubber bands of about the same size but different thicknesses across a shoe box or container. Show students the box guitar. Then hold the box so the students cannot see which rubber bands you are plucking as you pluck the rubber bands and have students listen to the sounds produced. Make sure students know the rubber bands are stretched the same amount. Have children circle or color the person with whom they agree the most and explain why they agree with that person. Explain to students that they should select the person whose idea best matches their thinking, not whose features they like most. See pages xxviii–xxxiii in the introduction for techniques used to guide "science talk" related to the probe.

Related Ideas in *Benchmarks for Science Literacy* (AAAS 2009)

K–2 Motion

- Things that make sound vibrate.

Related Core Ideas in *A Framework for K–12 Science Education* (NRC 2012)

K–2 PS4.A: Wave Properties

- Sound can make matter vibrate, and vibrating matter can make sound.

3–5 PS4.A: Wave Properties

- Waves of the same type can differ in amplitude (height of the wave) and wavelength (spacing between wave peaks). Waves can add or cancel one another as they cross, depending on their relative phase.

Related *Next Generation Science Standards* (Achieve Inc. 2013)

Grade 1: Waves and Their Applications in Technologies for Information Transfer

- 1-PS4-1: Plan and conduct investigations to provide evidence that vibrating materials can make sound and that sound can make materials vibrate.
- 1-PS4-4: Use tools and materials to design and build a device that uses light or sound to solve the problem of communicating over a distance.

Grade 4: Waves and Their Applications in Technologies for Information Transfer

- 4-PS4-1: Develop a model of waves to describe patterns in terms of amplitude and wavelength and that waves can cause objects to move.

Related Research

- In a study by Watt and Russell (1990), children suggest that sound is produced because an object is made of plastic or rubber or because it is thick, thin, taut, or hard.

Mechanisms for sound production offered by children were dependent on what the sound-producing object was; their descriptions of how sound is produced from a rubber band were very different from the explanations proposed for a drum.

- Children's explanations of how sound is produced can be sorted into three groups that involve physical properties of sound-producing materials, such as thickness, hardness, and elasticity; the size of the force needed to produce the sound; and vibrations (Driver, Squires, Rushworth, and Wood-Robinson 1994).
- Reference to movement or vibration increases with age (Driver, Squires, Rushworth, and Wood-Robinson 1994).
- Studies suggest that children may be confused about the speed or size of vibration: Bigger vibrations were thought to be slower than small vibrations and, consequently, difficulties arose in discussing pitch and volume (Driver, Squires, Rushworth, and Wood-Robinson 1994).

Suggestions for Instruction and Assessment

- A probe designed to elicit the idea that sound is produced by vibrations, "Making Sound," can be adapted for this grade level (Keeley, Eberle, and Farrin 2005).
- This probe can be used as a P-E-O (Predict-Explain-Observe) probe by having students predict which rubber band would make the lowest sound when plucked (Keeley 2008). If their observations do not match their prediction, they can revise their answer and make a new claim that is supported with evidence from the investigation.
- Develop an operational understanding of pitch as low to high sounds before introducing the formal scientific term *pitch*. Build a word wall of sound-related terms such as *pitch* and *vibration*.

- If tuning forks are available, have students investigate the pitch and differences in vibration.
- Use examples other than rubber bands and stringed instruments to develop pitch-related ideas. For example, provide students with opportunities such as filling bottles with different levels of water and tapping them to compare the pitch of the different sounds produced.

Related NSTA Resources

American Association for the Advancement of Science (AAAS). 2001. *Atlas of science literacy.* (See "Waves," pp. 64–65.) New York: Oxford University Press.

Ashbrook, P. 2009. The early years: Hear that? *Science and Children* 46 (8): 19–20.

Konicek-Moran, R. 2008. *Everyday science mysteries: Stories for inquiry-based science teaching.* (See "The Neighborhood Telephone System," pp. 160–168.) Arlington, VA: NSTA Press.

Ritz, W. 2007. *A head start on science.* Arlington, VA: NSTA Press.

Robertson, W. 2003. *Sound: Stop faking it! Finally understanding science so you can teach it.* Arlington, VA: NSTA Press.

Snyder, R., and J. Johnson. 2010. Science shorts: Do you hear what Horton hears? *Science and Children* 48 (2): 68–70.

References

Achieve Inc. 2013. *Next generation science standards. www.nextgenscience.org/next-generation-science-standards.*

American Association for the Advancement of Science (AAAS). 2009. Benchmarks for science literacy online. *www.project2061.org/publications/bsl/online*

Driver, R., A. Squires, P. Rushworth, and V. Wood-Robinson. 1994. *Making sense of secondary science: Research into children's ideas.* London: Routledge.

Keeley, P. 2008. *Science formative assessment: 75 practical strategies for linking assessment, instruction, and learning.* Thousand Oaks, CA: Corwin Press and Arlington, VA: NSTA Press.

Keeley, P., F. Eberle, and L. Farrin. 2005. *Uncovering student ideas in science, vol. 1: 25 formative assessment probes.* Arlington, VA: NSTA Press.

National Research Council (NRC). 2012. *A framework for K–12 science education: Practices, crosscutting concepts, and core ideas.* Washington, DC: National Academies Press.

Watt, D., and T. Russell. 1990. *Sound.* Primary SPACE Project Research Report. Liverpool, England: Liverpool University Press.

Big and Small Magnets

What are you thinking?

Big and Small Magnets

Teacher Notes

Purpose

The purpose of this assessment probe is to elicit children's ideas about magnets. The probe is designed to reveal children's ideas about the strength of a magnet in relation to the size of the magnet.

Related Concepts

magnetism, magnets

Explanation

Amir has the best answer: "Strength depends on the type of magnet." Sometimes the size of a magnet does not necessarily indicate its strength. For example, if you had two different sizes of magnets made of different materials, one may be stronger than the other. A small, powerful neodymium disc magnet could pick up a longer chain of paper clips than a larger ceramic disc magnet. Conversely, large magnets can be stronger than smaller magnets of the same type and material. For example, larger magnets inside a speaker are stronger than the smaller magnets in a speaker that are the same type of magnet and material. Magnets of any size can lose strength from heat, improper storage, or shock, such as hitting it with a hammer, and they can be weaker than a smaller magnet that has retained its strength.

Curricular and Instructional Considerations for Grades K–2

Investigating properties of magnets and materials attracted to magnets is a common activity in the primary grades. While details about magnetic force are developed in grades 3–5, investigating the properties of different magnets, such as comparing the strength of different types of magnets, is appropriate at this level.

Administering the Probe

Show students a variety of magnets of different types, sizes, and shapes. Ask students how the strength of the magnets can be determined or guide them to explain how a chain of paper clips could be used to determine the strength of the magnet (e.g., stronger magnets can hold a longer chain). Ask students if they think the bigger magnets are stronger than the smaller magnets. Go over each of the answer choices with them. Have children circle or color the person with whom they agree the most and explain why they agree with that person. Explain to students that they should select the person whose idea best matches their thinking, not whose features they like most. See pages xxviii–xxxiii in the introduction for techniques used to guide "science talk" related to the probe.

Related Ideas in *Benchmarks for Science Literacy* (AAAS 2009)

K–2 Forces of Nature

*   Magnets can be used to make some things move without being touched.

3–5 Forces of Nature

*   Without touching them, a magnet pulls on all things made of iron and either pushes or pulls on other magnets.

Related Core Ideas in *A Framework for K–12 Science Education* (NRC 2012)

K–2 PS.2.B: Types of Interactions

- When objects touch or collide, they push on one another and can change motion or shape.

3–5 PS.2.B: Types of Interactions

- Electric, magnetic, and gravitational forces between a pair of objects do not require that the objects be in contact—for example, magnets push or pull at a distance. The sizes of the forces in each situation depend on the properties of the objects and their distances apart and, for forces between two magnets, on their orientation relative to each other.

Related *Next Generation Science Standards* (Achieve Inc. 2013)

Kindergarten: Motion and Stability: Forces and Interactions

- K-PS2-1: Plan and conduct an investigation to compare the effects of different strengths or different directions of pushes and pulls on the motion of an object.

Grade 3: Motion and Stability: Forces and Interactions

- 3-PS2-3: Ask questions to determine cause and effect relationships of electric or magnetic interactions between two objects not in contact with each other.

Related Research

- Students may use an intuitive rule called "More A–More B" to reason why they think larger magnets are stronger. The larger the magnet, the stronger it is (Stavy and Tirosh 2000).
- A study by Finley (1986) looked at the effect of a television program on children's ideas about magnets. Before viewing the program, children thought big magnets were stronger than smaller ones. The effect of watching the television program that had a small, powerful magnet was simply to reverse their initial idea such that students then thought smaller magnets were stronger (Driver, Squires, Rushworth, and Wood-Robinson 1994).

Suggestions for Instruction and Assessment

- The probe "Magnets in Water" can be adapted for this grade level (Keeley and Tugel 2009).
- This probe can be used as a P-E-O (Predict-Explain-Observe) probe by having students make a claim (prediction) and then test the claim using a variety of magnets of different strengths and sizes and a chain of paper clips (Keeley 2008). If the observations do not match the prediction, students can make a new claim and support it with evidence from the investigation. **Safety note: Make sure children do not put small magnets in their mouth.**

Related NSTA Resources

Ashbrook, P. 2005. The early years: More than messing around with magnets. *Science and Children* 43 (2): 20–23.

Konicek-Moran, R. 2009. *More everyday science mysteries: Stories for inquiry-based science teaching.* (See "The Magnet Derby," pp. 149–158.) Arlington, VA: NSTA Press.

Kur, J., and M. Heitzmann. 2008. Attracting student wonderings. *Science and Children* 45 (5): 28–32.

References

Achieve Inc. 2013. *Next generation science standards.* *www.nextgenscience.org/next-generation-science-standards.*

American Association for the Advancement of Science (AAAS). 2009. Benchmarks for science literacy online. *www.project2061.org/publications/bsl/online*

Driver, R., A. Squires, P. Rushworth, and V. Wood-Robinson. 1994. *Making sense of secondary science: Research into children's ideas.* London: Routledge.

Finley, F. 1986. Evaluating instruction: The complementary use of clinical interviews. *Journal of Research in Science Teaching* 23 (17): 635–650.

Keeley, P. 2008. *Science formative assessment: 75 practical strategies for linking assessment, instruction, and learning.* Thousand Oaks, CA: Corwin Press and Arlington, VA: NSTA Press.

Keeley, P., and J. Tugel. 2009. *Uncovering student ideas in science, vol. 4: 25 new formative assessment probes.* Arlington, VA: NSTA Press.

National Research Council (NRC). 2012. *A framework for K–12 science education: Practices, crosscutting concepts, and core ideas.* Washington, DC: National Academies Press.

Stavy, R., and D. Tirosh. 2000. *How students (mis)understand science and mathematics: Intuitive rules.* New York: Teachers College Press.

Section 3

Earth and Space Science

Concept Matrix: Earth and Space Science
Probes #20–#25

PROBES / RELATED CONCEPTS ↓	20. What Makes up a Mountain?	21. Describing Soil	22. Is a Brick a Rock?	23. When Is My Shadow the Longest?	24. What Lights up the Moon?	25. When Is the Next Full Moon?
Earth materials	✓	✓	✓			
Earth-Sun system				✓		
human-made materials			✓			
lunar cycle						✓
Moon					✓	✓
Moonlight					✓	
Moon phases						✓
mountains	✓					
reflection					✓	
rock	✓		✓			
shadows				✓		
soil		✓				
Sun-Earth-Moon system					✓	✓

What Makes up a Mountain?

What are you thinking?

What Makes up a Mountain?

Teacher Notes

Purpose

The purpose of this assessment probe is to elicit children's ideas about the Earth materials that make up landforms. The probe is designed to reveal whether students recognize that mountains are made primarily of rock—a precursor idea to understanding the weathering of mountains by water, wind, and ice.

Related Concepts

mountains, Earth materials, rock

Explanation

Kelly has the best idea: "Mountains are made mostly of rock." Mountains are formed through volcanism or tectonic forces and are composed primarily of rock. For example, some mountains are formed when molten rock (magma) deep within the Earth makes its way to the surface and piles up to form a mountain—Mount St. Helens is an example. Some mountains are formed when tectonic plates, which are made of rock, collide and lead to folding of the Earth's crust. The Himalayas were formed in this way. Sometimes blocks of rock material slide along faults in the Earth's crust. The Sierra Nevada mountains were formed in this way. Mountains form over a long period of time, and the way they look can vary. However, if you could look below the surface, you would see mostly rock. Materials and organisms such as soil, trees and other plants, ice, and snow cover the rock on some parts of mountains. Some mountains have visible rock surfaces and outcroppings.

Curricular and Instructional Considerations for Grades K–2

In the early grades, children become familiar with and describe a variety of landforms, particularly ones that are found in their local environment, such as mountains, hills, valleys, cliffs, beaches, canyons, and caves. At this level, they are not expected to know how landforms are formed. Instead, the focus is on how landforms are shaped and changed by wind, water, and ice. At this level, students learn how materials such as the rock that makes up mountains can be worn down by wind, water, and ice into tiny particles over long periods of time. However, first they must know what materials make up landforms.

Administering the Probe

Show students a picture of a mountain or point out a familiar mountain in their local area as you use this probe. Encourage students to think below the trees, snow, and other things they can see on the mountain to imagine what mostly makes up the whole mountain. Have children circle or color the person with whom they agree the most and explain why they agree with that person. Explain to students that they should select the person whose idea best matches their thinking, not whose features they like most. See pages xxviii–xxxiii in the introduction for techniques used to guide "science talk" related to the probe.

Related Ideas in *Benchmarks for Science Literacy* (AAAS 2009)

K–2 Processes That Shape the Earth

- Chunks of rocks come in many sizes and shapes, from boulders to grains of sand and even smaller.

3–5 Processes That Shape the Earth

- Waves, wind, water, and ice shape and reshape the Earth's land surface by eroding rock and soil in some areas and depositing them in other areas, sometimes in seasonal layers.

Related Core Ideas in *A Framework for K–12 Science Education* (NRC 2012)

K–2 ESS1.C: The History of Planet Earth

- Some events, like an earthquake, happen very quickly; others, such as the formation of the Grand Canyon, occur very slowly, over a period of time much longer than one can observe.

3–5 ESS1.C: The History of Planet Earth

- Earth has changed over time. Understanding how landforms develop, are weathered, and erode can help infer the history of the current landscape.

Related *Next Generation Science Standards* (Achieve Inc. 2013)

Grade 2: Earth's Systems

- 2-ESS2-2: Develop a model to represent the shapes and kinds of land and bodies of water in an area.

Grade 4: Earth's Systems

- 4-ESS2-1: Make observations and/or measurements to provide evidence of the effects of weathering or the rate of erosion by water, ice, wind, or vegetation.

Related Research

In a study by Happs (1982), children described mountains as "high rocks" or "clumps of dirt or soil." A few children thought mountains were made of molten rock or "rock pushed up," while others thought all mountains are volcanoes (Driver, Squires, Rushworth, and Wood-Robinson 1994).

Suggestions for Instruction and Assessment

- Consider providing students with a set of labeled picture cards with various objects that are made of rock and not made of rock, including a card with the picture and label of a mountain. Using the card-sort strategy, have students sort the cards into things made of rock and things not made of rock. As students sort, look for where they place the mountain card and probe further by asking them why they think the mountain is (or is not) made of rock.
- Keep in mind that children in some geographic locales may have never seen an actual mountain. Others who do live near mountains may have observed only the external features that cover mountains, such as trees or snow. Provide a variety of photos or use video clips of different mountains and have children describe the mountains. How are they alike? How are they different? What do they see that is part of the mountain?
- Show students pictures or videos of mountains with visible rock faces and outcrops and ask them if they think the rest of the mountain is made mostly of rock material. Probe further to find out why they think

the rest of the mountain is or is not made of rock.

- Find photos of road cuts through mountains or tunnels such as the Allegheny Mountain Tunnel. If you Google images for mountain tunnels, you will find a variety of photos that show tunnels cut through sheer rock. This also adds an engineering connection in which students can see how engineers solve the problem of building roads in areas where there are tall mountain passes.

Related NSTA Resources

Konicek-Moran, R. 2010. *Even more everyday science mysteries: Stories for inquiry-based science teaching.* (See "A Day on Bare Mountain," pp. 71–81.) Arlington, VA: NSTA Press.

References

Achieve Inc. 2013. *Next generation science standards.* *www.nextgenscience.org/next-generation-science-standards.*

American Association for the Advancement of Science (AAAS). 2009. Benchmarks for science literacy online. *www.project2061.org/publications/bsl/online*

Driver, R., A. Squires, P. Rushworth, and V. Wood-Robinson. 1994. *Making sense of secondary science: Research into children's ideas.* London: Routledge.

Happs, J. 1982. *Rocks and minerals.* LISP Working Paper 204. Science Education Research Unit. University of Waikato, Hamilton, New Zealand.

National Research Council (NRC). 2012. *A framework for K–12 science education: Practices, crosscutting concepts, and core ideas.* Washington, DC: National Academies Press.

Describing Soil

What are you thinking?

Describing Soil

Teacher Notes

Purpose

The purpose of this assessment probe is to elicit children's ideas about soil. The probe is designed to find out if students recognize that soil is made up of living and nonliving material.

Related Concepts

soil, Earth materials

Explanation

Molly has the best answer: "Soil has living and nonliving parts." Soil is the solid material on the surface of Earth naturally formed through the interaction of physical elements and processes (e.g., weathering of rock by ice, water, and wind; erosion; and deposition) and biological activities (e.g., decay of dead organisms). Soil is a mixture made up of nonliving elements—such as water, minerals, and air—and living or once-living organisms or parts of organisms. The mineral part of soil comes from the weathering of rocks into tiny mineral particles. The once-living part of soil comes from decaying matter such as dead leaves. There are also living things in soil such as bacteria and fungi. There are many different types of soils and different ways to classify them, but all soil types can be described as having varying amounts of silt, sand, clay, and organic material.

Curricular and Instructional Considerations for Grades K–2

In the early grades, students should have opportunities to describe the properties of soil, observe different types of soils, and examine the composition of soil with magnifying

devices. Early experiences with soils help prepare students for learning about the weathering of rock and the interactions of organisms with their physical environment and connecting these ideas to the composition of soil.

Administering the Probe

Show students a sample of soil, but refrain from having them look closely or with magnifiers—this will come after they share their ideas about the probe. Have children circle or color the person with whom they agree the most and explain why they agree with that person. Explain to students that they should select the person whose idea best matches their thinking, not whose features they like most. See pages xxviii–xxxiii in the introduction for techniques used to guide "science talk" related to the probe.

Related Ideas in *Benchmarks for Science Literacy* (AAAS 2009)

K–2 Processes That Shape the Earth
- Chunks of rocks come in many sizes and shapes, from boulders to grains of sand and even smaller.

3–5 Processes That Shape the Earth
- Rock is composed of different combinations of minerals. Smaller rocks come from the breakage and weathering of bedrock and larger rocks. Soil is made partly from weathered rock, partly from plant remains—and also contains many living organisms.

Related Core Ideas in *A Framework for K–12 Science Education* (NRC 2012)

. . . .

K–2 ESS2.A: Earth Materials and Systems

- Wind and water can change the shape of the land. The resulting landforms, together with the materials on the land, provide homes for living things.

3–5 ESS2.A: Earth Materials and Systems

- Earth's major systems are the geosphere (solid and molten rock, soil, and sediments), the hydrosphere (water and ice), the atmosphere (air), and the biosphere (living things, including humans). These systems interact in multiple ways to affect Earth's surface materials and processes. Water, ice, wind, living organisms, and gravity break rocks, soils, and sediments into smaller particles and move them around.

Related *Next Generation Science Standards* (Achieve Inc. 2013)

. . . .

Kindergarten: Earth and Human Activity

- K-ESS3-1: Use a model to represent the relationship between the needs of different plants or animals (including humans) and the places they live.

Grade 5: Ecosystems: Interactions, Energy, and Dynamics

- 5-LS2-1: Develop a model to describe the movement of matter among plants, animals, decomposers, and the environment.

Grade 5: Earth's Systems

- 5-ESS2-1: Develop a model using an example to describe ways the geosphere, biosphere, hydrosphere, and/or atmosphere interact.

Related Research

- A study by Happs (1982) found that one of the most common misconceptions about the nature of soil was that soil is "just dirt" or "any stuff in the ground." Children seemed to be largely unaware of the role of living organisms or the identity of these organisms in the soil. Some children distinguished "dirt" from soil by saying soil has "more goodness in it" (Driver, Squires, Rushworth, and Wood-Robinson 1994).
- Research questions that asked students about the "disappearance" of dead animals or fruits on the surface of soil revealed several misconceptions about decay. Young children think dead things just disappear and no longer exist, not accounting for their becoming part of the soil. Some students recognized that rotted material "fertilizes" the soil but did not recognize it as being part of the soil. Generally, children are unaware of the role micro-organisms play in soil (Driver, Squires, Rushworth, and Wood-Robinson 1994).

Suggestions for Instruction and Assessment

- This probe can be extended using the card-sort strategy (Keeley 2008). Use printed pictures and words on cards and ask students to sort the cards into things that are part of soil and things that are not part of soil. You can include water, air, tiny bits of rock, micro-organisms (use only if children know what these are), clouds, trees, rotted leaves, earthworms, rotted animals, big rocks, glass, and plastic as examples.

- Provide children with magnifiers to examine local, naturally formed soil (not commercially sold potting soil) so they can see that soil is made of different parts.
- Provide children with different samples of soil to examine. Encourage them to describe how they are similar and different.
- Rub two rocks together to show how rock can be broken down (a process called weathering) into tiny particles and that these particles become soil.
- Have students build soil chambers in which they can bury plant material and observe how it decays and becomes part of the soil over time.
- Help students distinguish between the words *dirt* and *soil*. Soil is the useful material in which plants, including our food, grow; provides a place where some animals live; and is the surface that supports our buildings and roads. Dirt is soil in places where humans do not want it—for example, dirt on the floor or on your clothes from playing outside or mud on your shoes.

Related NSTA Resources

Gibb, L. 2000. Second-grade soil scientists. *Science and Children* 38 (3): 25–28.

Mayes, V. 2009. Natural resources: Digging soil. *Science and Children* 47 (3): 44–45.

NSTA Press. 2001. *Dig in! Hands-on soil investigations*. Arlington, VA: NSTA Press.

Piotrowski, J., T. Mildenstein, K. Dungan, and C. Brewer. 2007. The radish party. *Science and Children* 45 (2): 41–45.

References

Achieve Inc. 2013. *Next generation science standards. www.nextgenscience.org/next-generation-science-standards.*

American Association for the Advancement of Science (AAAS). 2009. Benchmarks for science literacy online. *www.project2061.org/publications/bsl/online*

Driver, R., A. Squires, P. Rushworth, and V. Wood-Robinson. 1994. *Making sense of secondary science: Research into children's ideas.* London: Routledge.

Happs, J. 1982. Some aspects of student understanding of soil. *Australian Science Teachers Journal* 28 (3): 25–31.

Keeley, P. 2008. *Science formative assessment: 75 practical strategies for linking assessment, instruction, and learning.* Thousand Oaks, CA: Corwin Press and Arlington, VA: NSTA Press.

National Research Council (NRC). 2012. *A framework for K–12 science education: Practices, crosscutting concepts, and core ideas.* Washington, DC: National Academies Press.

Is a Brick a Rock?

What are you thinking?

Is a Brick a Rock?

Teacher Notes

Purpose

The purpose of this assessment probe is to elicit children's ideas about rocks. The probe is designed to determine if children can distinguish between rocklike materials made by humans and rocks that have a geologic origin.

Related Concepts

rocks, Earth materials, human-made materials

Explanation

Chase has the better answer: "A brick is not a rock." Bricks are objects made by humans from Earth materials such as clay, lime and sand, or dried mud. They are often fired in a kiln. Rocks are a solid mass of mineral (a naturally occurring inorganic substance with a definite chemical composition) or mineral-like matter that occurs naturally on our planet and is formed by geologic processes. Rocks can be composed of one type of mineral or mixtures of two or more minerals. Bricks are also mixtures containing rock material, but they do not form naturally. Although bricks may contain some rock material, they are reshaped and recombined through a human manufacturing process, not a geologic process.

Curricular and Instructional Considerations for Grades K–2

Observing properties and classifying objects by their properties is an important part of the elementary science curriculum. Young children should become familiar with a variety of natural and human-made materials in their environment, describe the materials' properties, and classify them as natural or human made. They should develop an understanding that humans need materials from their environment and many of the materials humans use to build houses, roads, and other things come from the Earth. The focus is on natural resources used to make materials or build objects. At the grades 3–5 level, the focus on natural resources begins to move toward energy resources.

Administering the Probe

Ask children if they have ever seen a brick or things made from bricks. Show children an actual brick or pictures of structures made from bricks before they share their ideas related to the probe. Have children circle or color the person with whom they agree the most and explain why they agree with that person. Explain to students that they should select the person whose idea best matches their thinking, not whose features they like most. See pages xxviii–xxxiii in the introduction for techniques used to guide "science talk" related to the probe.

Related Ideas in *Benchmarks for Science Literacy* (AAAS 2009)

K–2 Structure of Matter

- Objects can be described in terms of their properties. Some properties, such as hardness and flexibility, depend upon what material the object is made of, and some properties, such as size and shape, do not.

3–5 Processes That Shape the Earth

- Rock is composed of different combinations of minerals.

3–5 Materials and Manufacturing

- Humans have produced a wide variety of materials, such as steel, plastic, and nylon, that do not appear in nature.

Related Core Ideas in *A Framework for K–12 Science Education* (NRC 2012)

. . . .

K–2 ESS3.A: Natural Resources

- Humans use natural resources for everything they do.

3–5 ESS3.A: Natural Resources

- All materials, energy, and fuels that humans use are derived from natural sources.

Related *Next Generation Science Standards* (Achieve Inc. 2013)

. . . .

Kindergarten: Earth and Human Activity

- K-ESS3-1: Use a model to represent the relationship between the needs of different plants or animals (including humans) and the places they live.

Grade 4: Earth and Human Activity

- 4-ESS3-1: Obtain and combine information to describe that energy and fuels are derived from natural resources and their uses affect the environment.

Related Research

- The word *rock* is used in many different ways, contributing to the confusion of what it means in a geologic sense (Freyberg 1985).
- In studies by Happs (1982, 1985), students had difficulty making the distinction between natural things and those created or altered by humans. For example,

some students considered a brick to be a rock because part of it comes from natural material (Driver, Squires, Rushworth, and Wood-Robinson 1994).

Suggestions for Instruction and Assessment

- A related probe, "Is It a Rock? Version 2," can be modified for use at this grade level (Keeley, Eberle, and Tugel 2007).
- This probe uses a format called "Opposing Views" and can also be used with a strategy called "Lines of Argumentation" (Keeley, in press). Have children sit in two lines that face each other—one line for those who selected Olivia and another line for those who selected Chase. Have students engage in argumentation—making a claim and justifying their claim with reasoning for why they think a brick is or is not a rock. If an argument is compelling enough to change a student's thinking, he or she can move to the other line.
- Take time to elicit students' ideas about what they think a rock is. Develop an operational definition before introducing the scientific definition of *rock*.
- Compare and contrast naturally formed rocks with objects that are "rocklike" and formed by humans (e.g., bricks, tiles, cement, clay pots), as well as objects that are shaped from rock (e.g., figurines carved out of rock, granite countertops, gravestones, statues).
- Use this probe as a nice tie-in to the *T* in STEM: technology. Science deals with our natural world (rocks); technology uses the application of science to modify our natural world to fulfill a human need (making clay pots out of rock). Furthermore, you can tie in the *E* (engineering) to show how engineers solve problems to meet human needs through a design process. (How can

materials be combined to make a clay pot that is hard to break?)

Related NSTA Resources

Ansberry, K., and E. Morgan. 2012. Rock solid science. In *Teaching science through trade books,* ed. C. A. Royce, E. Morgan, and K. Ansberry, pp. 229–233. Arlington, VA: NSTA Press.

Brkich, K. 2012. Science shorts: Is concrete a rock? *Science and Children* 50 (4): 80–82.

Ritz, B. 2007. *A head start on science.* Arlington, VA: NSTA Press.

Varelas, M., and J. Benhart. 2004. Welcome to rock day. *Science and Children* 40 (1): 40–45.

References

Achieve Inc. 2013. *Next generation science standards. www.nextgenscience.org/next-generation-science-standards.*

American Association for the Advancement of Science (AAAS). 2009. Benchmarks for science literacy online. *www.project2061.org/publications/bsl/online*

Driver, R., A. Squires, P. Rushworth, and V. Wood-Robinson. 1994. *Making sense of secondary science: Research into children's ideas.* London: Routledge.

Freyberg, P. 1985. Implications across the curriculum. In *Learning in science,* ed. R. Osborne and P. Freyberg, pp. 125–135. Auckland, New Zealand: Heinemann.

Happs, J. 1982. *Rocks and minerals.* LISP Working Paper 204. Hamilton, New Zealand: University of Waikato, Science Education Research Unit.

Happs, J. 1985. Regression in learning outcomes: Some examples from Earth science. *European Journal of Science Education* 7 (4): 431–443.

Keeley, P. In press. *Science formative assessment: 50 more strategies for linking assessment, instruction, and learning.* Thousand Oaks, CA: Corwin Press.

Keeley, P., F. Eberle, and J. Tugel. 2007. *Uncovering student ideas in science, vol. 2: 25 more formative assessment probes.* Arlington, VA: NSTA Press.

National Research Council (NRC). 2012. *A framework for K–12 science education: Practices, crosscutting concepts, and core ideas.* Washington, DC: National Academies Press.

When Is My Shadow the Longest?

☐ early morning

☐ late morning

☐ noon

What are you thinking?

When Is My Shadow the Longest?

Teacher Notes

Purpose

The purpose of this assessment probe is to elicit children's ideas about shadows. The probe is designed to find out how students think shadows change from sunrise to noon.

Related Concepts

Earth-Sun system, shadows

Explanation

The best response is "Early morning." Morning shadows are longest right at sunrise when the Sun is low on the horizon. The angle at which sunlight strikes Earth changes as the Sun appears to move across the sky due to the Earth's rotation. In the early morning, the Sun is seen as low on the horizon and the angle between the Sun's rays and Earth's horizon is small. The shadow that results from blocking the Sun's rays is long. As the angle between Earth's surface and the Sun's rays that strike the Earth's surface increases throughout the morning, the shadow gets shorter. At noon, the time when the Sun is highest in the sky, the size of a shadow is the shortest. After noon, the shadow will begin increasing until it reaches its longest length right before sunset. (*Note:* Solar noon, the time when the Sun is highest in the sky, may be different from "clock noon" [12:00 p.m.], depending on geographic location.)

Curricular and Instructional Considerations for Grades K–2

Observing changes in shadows is an appropriate activity to help primary-age children engage in scientific practices and identify patterns related to the Sun-Earth system. By observing and measuring shadows, students can describe the changing position of the Sun in relation to the Earth throughout the day. In addition to collecting data by observing changes in their shadows on a particular day, students can also collect and analyze data to describe seasonal changes to their shadows.

Administering the Probe

Adapt the probe to fit the hours of your school day. For example, for early morning, have students observe their shadow soon after they arrive at school (e.g., 8:30 a.m.). Wait about two hours before making a late-morning observation (e.g., 10:30 a.m.). List the time for students. Observe again at noon or close to noon. Make sure there is enough time in between to observe or measure the noticeable changes in shadow length. See pages xxviii–xxxiii in the introduction for techniques used to guide "science talk" related to the probe.

Related Ideas in *Benchmarks for Science Literacy* (AAAS 2009)

K–2 The Universe

- The Sun, Moon, and stars all appear to move slowly across the sky.

Related Core Ideas in *A Framework for K–12 Science Education* (NRC 2012)

K–2 ESS1.A: The Universe and Its Stars

- Patterns of the motion of the Sun, Moon, and stars in the sky can be observed, described, and predicted.

K–2 ESS1.B: Earth and the Solar System

- Seasonal patterns of sunrise and sunset can be observed, described, and predicted.

3–5 ESS1.B: Earth and the Solar System

- The orbits of Earth around the Sun and of the Moon around Earth, together with the rotation of Earth about an axis between its North and South poles, cause observable patterns. These include day and night; daily and seasonal changes in the length and direction of shadows; phases of the Moon; and different positions of the Sun, Moon, and stars at different times of the day, month, and year.

Related *Next Generation Science Standards* (Achieve Inc. 2013)

Grade 1: Earth's Place in the Universe

- 1-ESS1-1: Use observations of the Sun, Moon, and stars to describe patterns that can be predicted.

Grade 5: Earth's Place in the Universe

- 5-ESS1-2: Represent data in graphical displays to reveal patterns of daily changes in length and direction of shadows, day and night, and the seasonal appearance of some stars in the night sky.

Related Research

- Some researchers have found that children expect the shadow of an object to be the same shape as the object (Driver, Squires, Rushworth, and Wood-Robinson 1994).
- Students seem to have more success in locating where an object's shadow will fall in relation to a light source if the object is a person. They have more difficulty anticipating where a shadow will fall if it is a nonhuman object, such as a tree (Driver, Squires, Rushworth, and Wood-Robinson 1994).
- Plummer and Krajcik (2010) found that children as young as first grade knew that the Sun gets higher in the sky during the day and lower in the sky toward evening, although most were not able to accurately describe the Sun's path. Some even thought the Sun stopped moving during the day.

Suggestions for Instruction and Assessment

- A related probe, "Me and My Shadow," can be adapted for use with primary-grade students (Keeley, Eberle, and Dorsey 2008).
- This probe can be used to launch into inquiry after students have made their predictions about when their shadow would be the longest. Have students observe and measure their shadow throughout the morning until noon and describe the pattern.
- After students have discovered the pattern, relate the pattern to the position of the Sun throughout the day. However, use caution: **Never** allow children to look directly at the Sun.
- Make sure students know what a shadow is before asking them when their shadow will be longest.
- Have students model what happens to a shadow when the position of a light source—and thus the angle at which light strikes an object—changes. Provide

students with a flashlight and an upright object to test their ideas and record observations. Help students link their flashlight findings to the position of the Sun throughout the day.

- Extend the probe to include afternoon shadows. Have students draw a sequence of pictures to show the relationship between a shadow's length and the position of the Sun throughout the day.
- Extend shadow investigations across the school year to show there are seasonal patterns as well as daily patterns.

Related NSTA Resources

Ansberry, K., and E. Morgan. 2009. Teaching through trade books: Sunrise, sunset. *Science and Children* 46 (8): 14–16.

Barrows, L. 2007. Bringing light onto shadows. *Science and Children* 44 (9): 43–45.

Konicek-Moran, R. 2008. *Everyday science mysteries: Stories for inquiry-based science teaching.* (See "Where Are the Acorns?" pp. 39–50.) Arlington, VA: NSTA Press.

Konicek-Moran, R. 2011. *Even more everyday science mysteries: Stories for inquiry-based science teaching.* (See "Sunrise, Sunset," pp. 69–78.) Arlington, VA: NSTA Press.

Morgan, E. 2012. *Next time you see a sunset.* Arlington, VA: NSTA Press.

References

Achieve Inc. 2013. *Next generation science standards.* www.nextgenscience.org/next-generation-science-standards.

American Association for the Advancement of Science (AAAS). 2009. Benchmarks for science literacy online. *www.project2061.org/publications/bsl/online*

Driver, R., A. Squires, P. Rushworth, and V. Wood-Robinson. 1994. *Making sense of secondary science: Research into children's ideas.* London: Routledge.

Keeley, P., F. Eberle, and C. Dorsey. 2008. *Uncovering student ideas in science, vol. 3: Another 25 formative assessment probes.* Arlington, VA: NSTA Press.

National Research Council (NRC). 2012. *A framework for K–12 science education: Practices, crosscutting concepts, and core ideas.* Washington, DC: National Academies Press.

Plummer, J., and J. Krajcik. 2010. Building a progression for celestial motion: Elementary levels from an Earth-bound perspective. *Journal of Research in Science Teaching* 47 (7): 768–787.

What Lights up the Moon?

What are you thinking?

What Lights up the Moon?

Teacher Notes

Purpose

The purpose of this assessment probe is to elicit children's ideas about the Moon's light. The probe is designed to uncover children's ideas about the relationship between the Sun and the Moon and reflected light.

Related Concepts

Moon, reflection, moonlight, Sun-Earth-Moon system

Explanation

Billy has the best answer: "The light comes from the Sun." When the Sun shines on the Moon, half of the Moon is illuminated due to the reflection of sunlight off the surface of the Moon. During a full Moon, we see the entire lit portion of the Moon. Moon phases occur because we see only a portion of the part of the Moon reflecting light as the Moon orbits the Earth. When there is a full Moon, we can see better at night. When there is no full Moon, there is less light reflected from the Moon toward Earth, and it is darker than when there is a full Moon.

Curricular and Instructional Considerations for Grades K–2

In the primary grades, students observe the Moon in the daytime and nighttime sky. They observe different phases and describe the monthly patterns of changes. At this level, students can be challenged to think about why we can see the Moon, connecting Moon observations to the idea that we can see objects when light illuminates them. The challenge for students is to think about where the light we see on the Moon comes from, connecting to early ideas about light reflection.

Administering the Probe

Show students the probe and explain that these children are looking at a full Moon at night. Explain that the children in the probe have different ideas about moonlight. Ask students if they have ever seen a full Moon. Have them describe the full Moon. Ask them to think about where the light came from that makes the full Moon look bright at night. Have children circle or color the person with whom they agree the most and explain why they agree with that person. Explain to students that they should select the person whose idea best matches their thinking, not whose features they like most. See pages xxviii–xxxiii in the introduction for techniques used to guide "science talk" related to the probe.

Related Ideas in *Benchmarks for Science Literacy* (AAAS 2009)

K–2 The Universe

- The Moon looks a little different every day but looks the same again about every four weeks.

3–5 Motion

- Light travels and tends to maintain its direction of motion until it interacts with an object or material. Light can be absorbed, redirected, bounced back, or allowed to pass through.

Related Core Ideas in *A Framework for K–12 Science Education* (NRC 2012)

K–2 PS4.B: Electromagnetic Radiation

- Objects can be seen only when light is available to illuminate them.

3–5 PS4.B: Electromagnetic Radiation

- A great deal of light travels through space to Earth from the Sun and from distant stars.

Related *Next Generation Science Standards* (Achieve Inc. 2013)

Grade 1: Waves and Their Applications in Technologies for Information Transfer

- 1-PS4-2: Make observations to construct an evidence-based account that objects can be seen only when illuminated.

Grade 4: Waves and Their Applications in Technologies for Information Transfer

- 4-PS4-2: Develop a model to describe that light reflecting from objects and entering the eye allows objects to be seen.

Related Research

Understanding the phases of the Moon is challenging for students. Before students can understand the phases of the Moon, they must first master the idea of a spherical Earth and understand the concept of light reflection and how the Moon gets its light from the Sun (AAAS 2009).

Suggestions for Instruction and Assessment

- Related probes that can be modified for this grade level include "Moonlight" (Keeley and Tugel 2009) and "Crescent Moon" (Keeley and Sneider 2011).
- Compare and contrast objects that emit their own light (Sun and stars) and objects that reflect light, and connect this to how we see the Moon.
- For students who believe the Moon creates its own light, use a model to help them give up their misconception in favor of the scientific idea. Shine a light on a white or light-colored ball in a dark room. Have students describe what they see. Have them examine the ball to decide if there is something in it causing it to light up. Turn off the light and have students describe the ball. Link what happens when the light shines on the ball to what happens when the Sun shines on the Moon.
- Some students have difficulty understanding the Sun shines on the Moon at night because they cannot see the Sun. Use a model to show how the Sun shines on the Moon at night.
- On days when the Moon is visible in the daytime sky, challenge students to explain why we can see the Moon.

Related NSTA Resources

Ansberry, K., and E. Morgan. 2008. Teaching through trade books: Moon phases and models. *Science and Children* 46 (1): 20–22.

Young, T., and M. Guy. 2008. The Moon's phases and the self shadow. *Science and Children* 46 (1): 30–35.

References

Achieve Inc. 2013. *Next generation science standards.* *www.nextgenscience.org/next-generation-science-standards.*

American Association for the Advancement of Science (AAAS). 2009. Benchmarks for science literacy online. *www.project2061.org/publications/bsl/online*

Keeley, P., and C. Sneider. 2012. *Uncovering student ideas in astronomy: 45 new formative assessment probes.* Arlington, VA: NSTA Press.

Keeley, P., and J. Tugel. 2009. *Uncovering student ideas in science, vol. 4: 25 new formative assessment probes.* Arlington, VA: NSTA Press.

National Research Council (NRC). 2012. *A framework for K–12 science education: Practices, crosscutting concepts, and core ideas.* Washington, DC: National Academies Press.

When Is the Next Full Moon?

What are you thinking?

When Is the Next Full Moon?

Teacher Notes

Purpose
The purpose of this assessment probe is to elicit children's ideas about the lunar cycle. The probe is designed to reveal how long students think it takes to complete a lunar cycle.

Related Concepts
lunar cycle, Moon phases, Moon, Sun-Earth-Moon system

Explanation
Sasha has the best answer: "In about a month." As the Moon orbits around the Earth, its appearance as seen from Earth changes (phases of the Moon). Starting with a full Moon, the Moon goes through the following phases in this order, as seen from the Northern Hemisphere: full Moon, waning gibbous Moon, last quarter Moon, waning crescent Moon, new Moon, waxing crescent Moon, first quarter Moon, and waxing gibbous Moon (these are names for points in the Moon's gradual change). The average length of time between a full Moon and when the full Moon appears again (a lunar cycle) is about 29.5 days.

Curricular and Instructional Considerations for Grades K–2
The experience of observing the Moon should begin as early as possible. At this grade level, students make observations of the Moon in the nighttime (and daytime) sky—drawing and recording changes as the moon goes through a lunar cycle. Explaining what causes the phases of the Moon should wait until middle school. At the K–2 level, the emphasis should be on observation and analysis, recognition, and explanation of the Moon's repeated pattern.

Administering the Probe
Make sure students know what a full Moon is and explain that the children in the picture are looking at a full Moon at night. Have children circle or color the person with whom they agree the most and explain why they agree with that person. Explain to students that they should select the person whose idea best matches their thinking, not whose features they like most. As students explain their reasoning, you might consider asking them to draw what they would see in the sky when looking at the Moon, between the time a full Moon appears and the time it appears again. Consider using this probe and starting Moon observations on the first night of a full Moon. See pages xxviii–xxxiii in the introduction for techniques used to guide "science talk" related to the probe.

Related Ideas in *Benchmarks for Science Literacy* (AAAS 2009)

K–2 The Universe
- The Moon looks a little different every day but looks the same again about every four weeks.

3–5 The Universe

- The Earth is one of several planets that orbit the Sun, and the Moon orbits around the Earth.

Related Core Ideas in *A Framework for K–12 Science Education* (NRC 2012)

. . . .

K–2 ESS1.A: The Universe and Its Stars

- Patterns of the motion of the Sun, Moon, and stars in the sky can be observed, described, and predicted.

3–5 ESS1.B: Earth and the Solar System

- The orbits of Earth around the Sun and of the Moon around Earth, together with the rotation of the Earth about an axis between its North and South poles, cause observable patterns. These include day and night; daily and seasonal changes in the length and direction of shadows; phases of the Moon; and different positions of the Sun, Moon, and stars at different times of the day, month, and year.

Related *Next Generation Science Standards* (Achieve Inc. 2013)

. . . .

Grade 1: Earth's Place in the Universe

- 1-ESS1-1: Use observations of the sun, moon, and stars to describe patterns that can be predicted.

Grade 5: Earth's Place in the Universe

- 5-ESS1-2: Represent data in graphical displays to reveal patterns of daily changes in length and direction of shadows, day and night, and the seasonal appearance of some stars in the night sky.

Middle School: Earth's Place in the Universe

- MS-ESS1-1: Develop and use a model of the Earth-Sun-Moon system to describe the cyclic patterns of lunar phases, eclipses of the Sun and Moon, and seasons.

Related Research

- In a study involving 1,213 students in grades 5 through college, only 42% of the entire sample knew the Moon takes about one month to orbit the Earth (Schoon 1992).
- Most of the research related to phases of the Moon centers on children's explanations of what causes the Moon phases. One of the most common explanations for the changing appearance of the Moon is that the Earth casts its shadow on the Moon (Driver, Squires, Rushworth, and Wood-Robinson 1994).

Suggestions for Instruction and Assessment

- Combine this probe with a related probe, "Objects in the Sky," to determine if students recognize that the Moon can sometimes be observed in the daytime (Keeley, Eberle, and Tugel 2007).
- This probe can be used as a P-E-O (Predict-Explain-Observe) probe on the date the full Moon appears (Keeley 2008). After students have made their prediction and explained their thinking, have them observe and record daily Moon phases until the full Moon appears again. If their observations do not match their prediction, provide students with an opportunity to make a new claim and support it with evidence from their observations.
- Show students photographs of a changing moon. Ask students to describe how the visible part of the Moon changes.
- Keep a daily bulletin board of Moon phases. After two complete lunar cycles,

have children look at and describe the evidence that the cycle repeats

- The focus should be on describing the changes from day to day and identifying the repeating pattern of Moon phases. Learning the names of the Moon phases is less important than noting the cyclic pattern.

- If students are unable to view the Moon on some evenings (or in the daytime sky), use websites such as Moon Connection (*www.moonconnection.com*) to view the current Moon phase.

- Parents who work with their children at night to observe Moon phases run into some frustration when the Moon is not visible during their viewing time. Suggest a website parents can use (such as *www.stardate.org*) to show Moon phases and Moon rise and set times or print times when the Moon will rise and set. It is important for the teacher to inform parents that the Moon rises and sets at different times throughout the lunar cycle, so they might not see it each night, but the important thing is to get outdoors and make observations when you can.

Related NSTA Resources

Hubbard, L. 2008. Bringing Moon phases down to Earth. *Science and Children* 46 (1): 40–41.

Konicek-Moran, R. 2008. *Everyday science mysteries: Stories for inquiry-based science teaching.* (See "Moon Tricks?" pp. 29–38.) Arlington, VA: NSTA Press.

Royce, C., E. Morgan, and K. Ansberry. 2012. *Teaching science through trade books.* (See "Moon Phases and Models," pp. 305-312.)

Trundle, K., S. Wilmore, and W. Smith. 2006. The Moon project. *Science and Children* 43 (6): 52–55.

References

Achieve Inc. 2013. *Next generation science standards. www.nextgenscience.org/next-generation-science-standards.*

American Association for the Advancement of Science (AAAS). 2009. Benchmarks for science literacy online. *www.project2061.org/publications/bsl/online*

Driver, R., A. Squires, P. Rushworth, and V. Wood-Robinson. 1994. *Making sense of secondary science: Research into children's ideas.* London: Routledge.

Keeley, P. 2008. *Science formative assessment: 75 practical strategies for linking assessment, instruction, and learning.* Thousand Oaks, CA: Corwin Press and Arlington, VA: NSTA Press.

Keeley, P., F. Eberle, and J. Tugel. 2007. *Uncovering student ideas in science, vol. 2: 25 more formative assessment probes.* Arlington, VA: NSTA Press.

National Research Council (NRC). 2012. *A framework for K–12 science education: Practices, crosscutting concepts, and core ideas.* Washington, DC: National Academies Press.

Schoon, K. J. 1992. Students' alternative conceptions of Earth and space. *Journal of Geological Education* 40: 209–214.

Index

Note: Page numbers in *italics* refer to charts.

Index

Index

Index

Index

Index

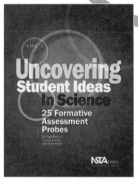

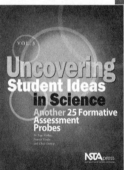
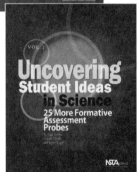

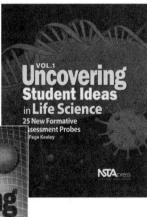